UCLA Symposia on Molecular and Cellular Biology, New Series

Series Editor, C. Fred Fox

RECENT TITLES

Volume 74
Growth Regulation of Cancer,
Marc E. Lippman, *Editor*

Volume 75
Steroid Hormone Action, Gordon Ringold, *Editor*

Volume 76
Molecular Biology of Intracellular Protein Sorting and Organelle Assembly,
Ralph A. Bradshaw, Lee McAlister-Henn, and Michael G. Douglas, *Editors*

Volume 77
Signal Transduction in Cytoplasmic Organization and Cell Motility, Peter Satir, John S. Condeelis, and Elias Lazarides, *Editors*

Volume 78
Tumor Progression and Metastasis,
Garth L. Nicolson and Isaiah J. Fidler, *Editors*

Volume 79
Altered Glycosylation in Tumor Cells,
Christopher L. Reading, Senitiroh Hakomori, and Donald M. Marcus, *Editors*

Volume 80
Protein Recognition of Immobilized Ligands,
T.W. Hutchens, *Editor*

Volume 81
Biological and Molecular Aspects of Atrial Factors, Philip Needleman, *Editor*

Volume 82
Oxy-Radicals in Molecular Biology and Pathology, Peter A. Cerutti, Irwin Fridovich, and Joe M. McCord, *Editors*

Volume 83
Mechanisms and Consequences of DNA Damage Processing, Errol C. Friedberg and Philip C. Hanawalt, *Editors*

Volume 84
Technological Advances in Vaccine Development, Laurence Lasky, *Editor*

Volume 85
B Cell Development, Owen N. Witte, Norman R. Klinman, and Maureen C. Howard, *Editors*

Volume 86
Synthetic Peptides: Approaches to Biological Problems, Emil Thomas Kaiser and James P. Tam, *Editors*

Volume 87
Gene Transfer and Gene Therapy, I. Verma, R. Mulligan, and A. Beaudet, *Editors*

Volume 88
Molecular Biology of the Eye: Genes, Vision, and Ocular Disease, Joram Piatigorsky, Toshimichi Shinohara, and Peggy S. Zelenka, *Editors*

Volume 89
Liposomes in the Therapy of Infectious Diseases and Cancer, Gabriel Lopez-Berestein and Isaiah J. Fidler, *Editors*

Volume 90
Cell Biology of Virus Entry, Replication, and Pathogenesis, Richard W. Compans, Ari Helenius, and Michael B.A. Oldstone, *Editors*

Volume 91
Bone Marrow Transplantation: Current Controversies, Richard E. Champlin and Robert Peter Gale, *Editors*

Volume 92
The Molecular Basis of Plant Development,
R. Goldberg, *Editor*

Volume 93
Cellular and Molecular Biology of Muscle Development, Laurence H. Kedes and Frank E. Stockdale, *Editors*

Volume 94
Molecular Biology of RNA, Thomas R. Cech, *Editor*

Volume 95
DNA-Protein Interactions in Transcription,
Jay D. Gralla, *Editor*

Volume 96
Stress-Induced Proteins, Mary Lou Pardue, James R. Feramisco, and Susan Lindquist, *Editors*

Volume 97
Molecular Biology of Stress, Shlomo Breznitz and Oren Zinder, *Editors*

Volume 98
Metal Ion Homeostasis: Molecular Biology and Chemistry, Dean H. Hamer and Dennis R. Winge, *Editors*

Volume 99
Human Tumor Antigens and Specific Tumor Therapy, Richard S. Metzgar and Malcolm S. Mitchell, *Editors*

Please contact the publisher for information about previous titles in this series.

UCLA Symposia Board

Stress-Induced Proteins

Stress-Induced Proteins

Proceedings of a Hoffmann-La Roche-Director's Sponsors-UCLA Symposium
Held at Keystone, Colorado
April 10–16, 1988

Editors

Mary Lou Pardue
Department of Biology
Massachusetts Institute of Technology
Cambridge, Massachusetts

James R. Feramisco
Cancer Center
University of California, San Diego,
La Jolla, California

Susan Lindquist
Departments of Molecular Genetics and Cell Biology
University of Chicago
Chicago, Illinois

Alan R. Liss, Inc. • New York

Address all Inquiries to the Publisher
Alan R. Liss, Inc., 41 East 11th Street, New York, NY 10003

Printed in the United States of America

Library of Congress Cataloging-in-Publication Data

Stress-induced proteins.

(UCLA symposia on molecular and cellular biology; new ser., v. 96)
"UCLA Symposium on Stress-Induced Proteins, held at Keystone, Colorado, April 10–16, 1988"—Pref.
Includes index.
1. Heat shock proteins—Congresses. 2. Gene expression—Congresses. I. Pardue, Mary Lou. II. Feramisco, James R. III. Lindquist, Susan. IV. UCLA Symposium on Stress-Induced Proteins (1988 : Keystone, Colo.) V. Series.
QP552.H43S77 1988 599′.01 88-13850

ISBN 0-8451-2695-4

Contents

VI. CLINICAL PROBLEMS

Contributors

Manohar K. Adwankar, Department of Radiation Oncology, Stanford University, Stanford, CA 94305; present address: Cancer Research Institute, Tata Memorial Center, Parel, Bombay 400 012, India **[223]**

Neus Agell, Department of Microbiology and Immunology, Washington University School of Medicine, St. Louis, MO 63110 **[137]**

Robin L. Anderson, Department of Radiation Oncology, Stanford University, Stanford, CA 94305 **[223]**

D. Ang, Department of Cellular, Viral, and Molecular Biology, University of Utah, Salt Lake City, UT 84132 **[37]**

A.-P. Arrigo, Department of Cell Biology, Cold Spring Harbor Laboratory, Cold Spring Harbor, NY 11724 **[187]**

Vathsala S. Basrur, Department of Radiation Oncology, Stanford University, Stanford, CA 94305 **[223]**

Etienne-Emile Baulieu, INSERM U 33, and Faculty de Medecine Paris-Sud, Lab. Hormones, 94275 Bicêtre Cedex, France **[203]**

W.G. Bendena, Department of Biology, Massachusetts Institute of Technology, Cambridge, MA 02139; present address: Department of Biology, Queen's University, Kingston, Ontario, Canada **[3]**

Günter Blobel, Laboratory of Cell Biology, Howard Hughes Medical Institute, The Rockefeller University, New York, NY 10021 **[163]**

Ursula Bond, Department of Microbiology and Immunology, Washington University School of Medicine, St. Louis, MO 63110; present address: Department of Molecular Biophysics and Biochemistry, Yale University, New Haven, CT 06510 **[137]**

J. José Bonner, Institute for Molecular and Cellular Biology, and Department of Biology, Indiana University, Bloomington, IN 47405 **[95]**

William Boorstein, Department of Physiological Chemistry, University of Wisconsin-Madison, Madison, WI 53706 **[51]**

Bernd Bukau, Department of Biology, Massachusetts Institute of Technology, Cambridge, MA 02139 **[27]**

The numbers in brackets are the opening page numbers of the contributors' articles.

Stephen W. Carper, Department of Radiation Oncology and Cancer Center, University of Arizona Health Sciences Center, Tuscon, AZ 85724 **[247]**

I.L. Cartwright, Department of Biology, Washington University, St. Louis, MO 63130; present address: Department of Biochemistry and Molecular Biology, University of Cincinnati College of Medicine, Cincinnati, OH 45267 **[15]**

Maria-Grazia Catelli, INSERM U 33, and Faculte de Medecine Paris-Sud, Lab. Hormones, 94275 Bicêtre Cidex, France **[203]**

G.N. Chandrasekhar, Department of Cellular, Viral, and Molecular Biology, University of Utah, Salt Lake City, UT 84132; present address: Department of Microbiology, Columbia University, New York, NY 10032 **[37]**

William J. Chirico, Laboratory of Cell Biology, Howard Hughes Medical Institute, The Rockefeller University, New York, NY 10021 **[163]**

Nancy C. Collier, Department of Microbiology and Immunology, Washington University School of Medicine, St. Louis, MO 63110 **[137]**

Elizabeth Craig, Department of Physiological Chemistry, University of Wisconsin-Madison, Madison, WI 53706 **[51]**

Q. Deveraux, Department of Biochemistry, University of Utah School of Medicine, Salt Lake City, UT 84132 **[149]**

T.J. Dietz, Department of Biology, Washington University, St. Louis, MO 63130 **[15]**

Caroline E. Donnelly, Department of Biology, Massachusetts Institute of Technology, Cambridge, MA 02139 **[27]**

S.C.R. Elgin, Department of Biology, Washington University, St. Louis, MO 63130 **[15]**

P.E. Falkenburg, Molekulare Genetik / ZMBH, University of Heidelberg, FRG-6900 Heidelberg, Federal Republic of Germany **[175]**

O. Fayet, Department of Cellular, Viral, and Molecular Biology, University of Utah, Salt Lake City, UT 84132; present address: CNRS, 31026 Toulouse, France **[37]**

James R. Feramisco, Cold Spring Harbor Laboratory, Cold Spring Harbor, NY 11724; Present address: Cancer Center, University of California, San Diego, La Jolla, CA 92023 **[xvii]**

M.E. Fini, Department of Biology, Massachusetts Institute of Technology, Cambridge, MA 02139 **[3]**

David J.M. Fuller, Department of Radiation Oncology and Cancer Center, University of Arizona Health Sciences Center, Tuscon, AZ 85724 **[247]**

J.C. Garbe, Department of Biology, Massachusetts Institute of Technology, Cambridge, MA 02139 **[3]**

C. Georgopoulos, Department of Cellular, Viral, and Molecular Biology, University of Utah, Salt Lake City, UT 84132 **[37]**

Eugene W. Gerner, Department of Radiation Oncology and Cancer Center, University of Arizona Health Sciences Center, Tuscon, AZ 85724 **[247]**

D.S. Gilmour, Department of Biology, Washington University, St. Louis, MO 63130 **[15]**

Ch. Haass, Molekulare Genetik/ ZMBH, University of Heidelberg, FRG-6900 Heidelberg, Federal Republic of Germany **[175]**

George M. Hahn, Department of Radiation Oncology, Stanford University, Stanford, CA 94305 **[223]**

Elizabeth M. Hallberg, Department of Zoology, Iowa State University, Ames, IA 50011 **[107]**

Richard L. Hallberg, Department of Zoology, Iowa State University, Ames, IA 50011 **[107]**

Mark Hallett, Institute for Molecular and Cellular Biology, and Department of Biology, Indiana University, Bloomington, IN 47405 **[95]**

Paul M. Harari, Department of Radiation Oncology and Cancer Center, University of Arizona Health Sciences Center, Tucson, AZ 85724 **[247]**

Rubén Henríquez, Laboratory of Cell Biology, Howard Hughes Medical Institute, The Rockefeller University, New York, NY 10021 **[163]**

Seiji Hongo, Department of Developmental Genetics and Anatomy, Case Western Reserve University, Cleveland, OH 44106 **[129]**

Marcelo Jacobs-Lorena, Department of Developmental Genetics and Anatomy, Case Western Reserve University, Cleveland, OH 44106 **[129]**

Kathleen John-Adler, Department of Pharmacology, UMDNJ-Robert Wood Johnson Medical School, Piscataway, NJ 08854 **[117]**

G.M. Kidder, Department of Biology, Massachusetts Institute of Technology, Cambridge, MA 02139; present address: Department of Zoology, University of Western Ontario, N6A 5B7 London, Ontario, Canada **[3]**

P.-M. Kloetzel, Molekulare Genetik/ ZMBH, University of Heidelberg, FRG-6900 Heidelberg, Federal Republic of Germany **[175]**

S.C. Lakhotia, Department of Biology, Massachusetts Institute of Technology, Cambridge, MA 02139; present address: Department of Zoology, Banaras Hindu University, 221 005 Varanasi, India **[3]**

Raju Lathigra, MRC Tuberculosis and Related Infections Unit, Royal Postgraduate Medical School, Hammersmith Hospital, London, W12 0HS, United Kingdom **[275]**

K. Liberek, Division of Biophysics, Department of Molecular Biology, University of Gdansk, 80-822 Gdansk, Poland **[37]**

Susan Lindquist, Departments of Molecular Genetics and Cell Biology, University of Chicago, Chicago, IL 60637 **[xvii]**

John T. Lis, Section of Biochemistry, Molecular and Cellular Biology, Cornell University, Ithaca, NY 14853 **[73]**

Kunihiro Matsumoto, Department of Cellular and Molecular Biology, DNAX Research Institute, Palo Alto, CA 94304-1104 **[63]**

Monica McAndrews, Institute for Molecular and Cellular Biology, and Department of Biology, Indiana University, Bloomington, IN 47405 **[95]**

Terrill K. McClanahan, Department of Biochemistry, Molecular Biology, and Cell Biology, Northwestern University, Evanston, IL 60208 **[83]**

Angela Mehlert, MRC Tuberculosis and Related Infections Unit, Royal Postgraduate Medical School, Hammersmith Hospital, London, W12 0HS, United Kingdom **[275]**

Herschel K. Mitchell, Division of Biology, California Institute of Technology, Pasadena, CA 91125 **[235]**

L.A. Mizzen, Department of Cell Biology, Cold Spring Harbor Laboratory, Cold Spring Harbor, NY 11724 **[187]**

Richard I. Morimoto, Department of Biochemistry, Molecular Biology, and Cell Biology, Northwestern University, Evanston, IL 60208 **[83]**

Dick Mosser, Department of Biochemistry, Molecular Biology, and Cell Biology, Northwestern University, Evanston, IL 60208 **[83]**

Charles Nicolet, Department of Physiological Chemistry, University of Wisconsin-Madison, Madison, WI 53706 **[51]**

Mary Lou Pardue, Department of Biology, Massachusetts Institute of Technology, Cambridge, MA 02139 **[xvii, 3]**

Hay-Oak Park, Department of Physiological Chemistry, University of Wisconsin-Madison, Madison, WI 53706 **[51]**

Janice Parker-Thornburg, Institute for Molecular and Cellular Biology, and Department of Biology, Indiana University, Bloomington, IN 47405 **[95]**

Olga Perisic, Section of Biochemistry, Molecular and Cellular Biology, Cornell University, Ithaca, NY 14853 **[73]**

David H. Perlmutter, Departments of Pediatrics, Cell Biology, and Physiology, Washington University School of Medicine, St. Louis, MO 63110 **[257]**

Nancy S. Petersen, Department of Molecular Biology, University of Wyoming, Laramie, WY 82071 **[235]**

G. Pratt, Department of Biochemistry, University of Utah School of Medicine, Salt Lake City, UT 84132 **[149]**

Su Qian, Department of Developmental Genetics and Anatomy, Case Western Reserve University, Cleveland, OH 44106 **[129]**

M. Rechsteiner, Department of Biochemistry, University of Utah School of Medicine, Salt Lake City, UT 84132 **[149]**

Marilyn M. Sanders, Department of Pharmacology, UMDNJ-Robert Wood Johnson Medical School, Piscataway, NJ 08854 **[117]**

Milton J. Schlesinger, Department of Microbiology and Immunology, Washington University School of Medicine, St. Louis, MO 63110 **[137]**

Ann C. Sherwood, Department of Pharmacology, UMDNJ-Robert Wood Johnson Medical School, Piscataway, NJ 08854 **[117]**

E. Siegfried, Department of Biology, Washington University, St. Louis, MO 63130 **[15]**

J. Spence, Department of Cellular, Viral, and Molecular Biology, University of Utah, Salt Lake City, UT 84132; present address: Department of Biology, Massachusetts Institute of Technology, Cambridge, MA 02139 **[37]**

David Stone, Department of Physiological Chemistry, University of Wisconsin-Madison, Madison, WI 53706 **[51]**

Kazuma Tanaka, Department of Fermentation Technology, Hiroshima University, Higashihiroshima 724, Japan **[63]**

Nicholas G. Theodorakis, Department of Biochemistry, Molecular Biology, and Cell Biology, Northwestern University, Evanston, IL 60208 **[83]**

G.H. Thomas, Department of Biology, Washington University, St. Louis, MO 63130 **[15]**

K. Tilly, Laboratory of Biochemistry, NCI-NIH, Bethesda, MD 20892 **[37]**

Akio Toh-e, Department of Fermentation Technology, Hiroshima University, Higashihiroshima 724, Japan **[63]**

Graham C. Walker, Department of Biology, Massachusetts Institute of Technology, Cambridge, MA 02139 **[27]**

M. Gerard Waters, Laboratory of Cell Biology, Howard Hughes Medical Institute, The Rockefeller University, New York, NY 10021 **[163]**

W.J. Welch, Department of Cell Biology, Cold Spring Harbor Laboratory, Cold Spring Harbor, NY 11724 **[187]**

Gregg Williams, Department of Biochemistry, Molecular Biology, and Cellular Biology, Northwestern University, Evanston, IL 60208 **[83]**

Hua Xiao, Section of Biochemistry, Molecular and Cellular Biology, Cornell University, Ithaca, NY 14853 **[73]**

Takehiro Yatomi, Department of Fermentation Technology, Hiroshima University, Higashihiroshima 724, Japan **[63]**

Douglas Young, MCR Tuberculosis and Related Infections Unit, Royal Postgraduate Medical School, Hammersmith Hospital, London, W12 0HS, United Kingdom **[275]**

T. Ziegelhoffer, Department of Cellular, Viral, and Molecular Biology, University of Utah, Salt Lake City, UT 84132 **[37]**

M. Zylicz, Division of Biophysics, Department of Molecular Biology, University of Gdansk, 80-822 Gdansk, Poland **[37]**

Preface

The heat-shock (stress) response is an important homeostatic mechanism that enables the cell to survive a variety of environmental stresses. The fundamental nature of this mechanism is underscored by the fact that the heat-shock response is found in all animals, plants, and bacteria. Study of the heat-shock response is a young research discipline that brought together 186 biologists for the UCLA Symposium on **Stress-Induced Proteins,** held at Keystone, Colorado, April 10–16, 1988. This was the third major meeting for this field of study, the first of which was held in May 1982 at the Cold Spring Harbor Laboratory, Cold Spring Harbor, New York.

The heat-shock response is essentially a single-cell response, and even for multicellular organisms, it can be studied with homogeneous populations of cultured cells, circumventing the complexities of interactions between different cell types. The experimental advantages of studying heat shock in cultured cells, in yeast, and in bacteria are leading to rapid advances toward understanding its molecular biology. However, these studies have left open important questions about the role of heat shock in the complex physiology of multicellular organisms. In the heat-shock response the individual cell is both the sensor and the responder to the stress. For multicellular animals, many stresses are perceived by a limited set of cells or tissues, such as the nervous system or the immune system, yet many other cell types take part in the response. What role does the heat-shock response play in this type of stress? As a way of beginning the discussions needed to explore these questions, the symposium was held concurrently with the UCLA Symposium on **The Molecular Biology of Stress,** at which the participants discussed whole-organismal types of stress. The two groups had two joint plenary sessions, one formal discussion, and many informal discussions.

The work presented at the **Stress-Induced Proteins** meeting covered many aspects of the molecular biology of the heat-shock response and several other related single-cell responses. These studies are advancing our understanding of both stressed and normal cells. Genes for the major heat-shock-induced proteins (HSPs) have been cloned and sequenced from many organisms. The proteins fall into a small number of highly conserved families. It is now apparent that all of these heat-shock proteins or related proteins are also expressed in nonstressed cells, at least at some times or in some tissues. Although the functions of these proteins in heat shock are not yet understood, reports at the symposium show that much has been learned about the functions of these proteins in normal cells.

Although the stress-induced proteins have attracted a great deal of attention, many cellular activities and structures are modified during the heat-shock response. New, noncoding RNAs are induced, RNA processing is altered, cytoskeletal rearrangements are induced, and other cell components are rearranged. In addition, protein turnover is altered. Papers in this volume report findings on the molecular bases and the physiological consequences of these phenomena.

The heat-shock response provides a small set of coordinately controlled genes whose transcription can be manipulated easily by the investigator. Work presented at this symposium sheds new light on the chromatin structure around such genes. A very exciting recent result has been the characterization of transcription factors from *Drosophila*, yeast, mammals, and bacteria.

The evolutionary conservation of the heat-shock response is strong evidence that the response is beneficial to cells. The most obvious advantage conferred by heat shock is thermotolerance. Induction of a low level of heat-shock synthesis allows the cell or the organism to withstand temperatures that would otherwise produce severe damage or death. The induced thermotolerance is transient and is lost over a period of hours or days. An understanding of thermotolerance has important practical applications, such as extending the growth range of economically important plants or preventing developmental abnormalities caused by high temperature. This importance has been recognized for some time in cancer therapy where hyperthermia is one of the weapons used to treat tumors. Recent work on the mechanisms of thermotolerance was another topic of the meeting.

A recent and unexpected finding is that stress proteins are major antigens in many infectious diseases. This relationship between the heat-shock response and the immune system emphasizes the fundamental nature of the response and its interaction with many aspects of the biology of the organism.

We are grateful for the support provided to the symposium by Hoffmann-La Roche, Inc., and the UCLA Symposia Director's Sponsors—Cetus Corporation, ICI Pharmaceuticals Group, Monsanto, Shering Corporation, and The Upjohn Company. In addition, we would like to thank Robin Yeaton and Betty Handy for the very smooth organization of the meeting.

Mary Lou Pardue
James R. Feramisco
Susan Lindquist

I. ORGANIZATION OF HEAT SHOCK GENES

Stress-Induced Proteins, pages 3–14

hsrω: A DIFFERENT SORT OF HEAT SHOCK LOCUS[1]

W.G.Bendena[2], M.E. Fini, J.C. Garbe, G.M. Kidder[3], S.C. Lakhotia[4], and M.L. Pardue

Department of Biology, Massachusetts Institute of Technology, Cambridge, MA 02139

ABSTRACT The hsrω locus is very unusual. Each Drosophila species has one hsrω locus, identified by its puffing phenotype. This identification is confirmed by molecular analysis. The hsrω locus produces 3 major transcripts and appears to have both nuclear and cytoplasmic functions.

INTRODUCTION

Cytological studies of the major heat shock puffs of Drosophila led directly to the identification and characterization of the major heat shock proteins. However, one of the largest and most actively transcribed heat shock puffs has remained without an assigned protein product. This puff is in polytene region 93D in D. melanogaster, 48B in D. hydei, 20CD in D. virilis, and 58C in D. pseudoobscura. In spite of the different polytene band numbers in the different species, cytological studies have suggested that these puffs are equivalent. The puffs respond not only to agents that induce the heat shock response, but also to some agents that affect only this puff. The puffs contain 300 nM RNP granules not seen in other puffs (reviewed by 1). Recently the cytological

---

1. This work was supported by a grant from the National Institutes of Health.
2. Present address: Queen's University, Kingston, Ontario
3. Present address: University of Western Ontario, London, Ontario
4 Present address: Banaras Hindu University, Varanasi, India

evidence of the equivalence of these puffs has been confirmed by analyses of DNA sequences (2; 3).

Our studies on the 93D locus and its relatives in the other species suggest that the important products of this locus are RNAs, not proteins. For this reason, 93D and the equivalent loci in the other Drosophila species have been designated the hsrω (for heat shock RNA omega) locus. In each species this locus produces three major RNAs, named ω1, ω2, and ω3. These RNAs make up a set that is very different from the transcripts of any known locus.

THE MULTIPLE hsrω TRANSCRIPTS APPEAR TO HAVE SEPARATE ROLES IN THE NUCLEUS AND THE CYTOPLASM

Analyses of RNA from many different cell types show the same three major hsrω transcripts in each cell type. These transcripts are produced constitutively in most cell types but the level of the transcripts is increased severalfold by heat shock (Figure 1, A and B). In this respect the hsrω locus resembles the hsp83 locus. The hsrω locus also shows some developmental control of its activity (Figure 1, D-E), thus resembling the hsp22-27 group (reviewed by 4).

When different Drosophila species are compared there are minor differences in the sizes of the three hsrω transcripts: however, in general, the ω1 RNA is larger than 9 kb, the ω2 RNA is approximately 2 kb, and the ω3 RNA is approximately 1.2 kb. The ω1 and ω2 transcripts are found only in the nucleus; ω3 is found in the nucleus and cytoplasm (Figure 2). All three transcripts have the same transcriptional start site; however there are several reasons for thinking that ω1 is not a precursor to ω2 and ω3. First, there is a polyadenylation site near the end of the ω2 sequence, suggesting that ω2 transcripts are produced by alternate termination of transcription in this region. Second, we do not see any of the RNA fragments that would be expected if the terminal regions of ω1 were cleaved off to produce ω2. Third, ω1 is much more abundant in the nucleus than any known precursors. Finally, the pattern of evolution of the sequences that are found only in ω1 argues that these sequences have a specific nuclear function (see below). The ω2 RNA differs from the ω3 RNA only by the removal of a 700 nt intron. Thus ω2 does appear to be the nuclear precursor to ω3 (2).

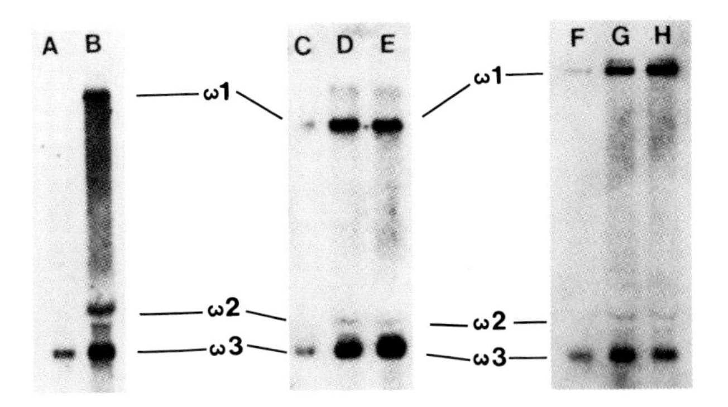

Figure 1. hsrω of D. melanogaster is heat shock inducible and developmentally regulated. Blot hybridization of control (A) and heat shock (B) RNA isolated from D. melanogaster embryos shows induction of the 3 major hsrω transcripts. Developmental increases in the level of hsrω transcripts are also detected. RNA from (non-heat shocked) early (C) middle (D) and late (E) stage embryos shows that hsrω transcripts increase during embryogenesis. Low levels of hsrω are detected during second instar (F) but increased levels are seen during third instar (G) and pupal (H) stages. Within each panel equal amounts of RNA were used for each lane. Blots were probed with a cDNA for hsrω3 to detect all three transcripts.

hsrω SEQUENCES ARE HIGHLY DIVERGENT BUT HAVE IMPORTANT CONSERVED FEATURES

The genes encoding heat shock proteins show strong DNA sequence conservation between animals, plants, and bacteria (4). In contrast, the hsrω loci have diverged even within the genus Drosophila. When small fragments of DNA from the D. melanogaster hsrω at 93D are hybridized to DNA from the D. hydei hsrω at 48B, the only significant cross-hybridization detected at any stringency is a fragment that includes the 3′ end of the intron and part of the second exon. The hybridization is due to 59 bp of perfect homology that begins 39 bp from the 3′ end of the intron. D. melanogaster and D. hydei are separated by 60 million years. When the sequence of D. melanogaster is compared with that of 58C from D. pseudoobscura, which is

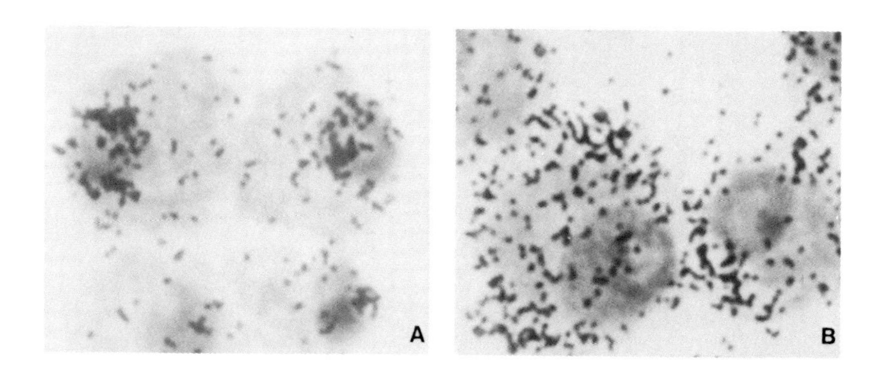

Figure 2. Autoradiograph of transcripts detected by in situ hybridization. Transcripts of hsrω show a different distribution from those of other heat shock genes. Cultured Schneider cells were heat shocked for one hour at 36°C, then spun onto slides, fixed, permeabilized, and hybridized to ^{3}H-RNA complementary to ω3 (A) or hsp70 (B). The ω3 probe detects all 3 hsrω transcripts. The hsrω transcripts are concentrated near the periphery of the nucleus with only a small fraction in the cytoplasm. hsp70 RNA is predominantly cytoplasmic with very little detected in the nucleus. Giemsa stain. Magnification x1850.

separated by 30 million years, the perfect homology has been extended by 2 bp in the intron. Cross-hybridization indicates that the conservation of this particular region extends throughout the genus (Garbe et al., in preparation).

The conservation of the intron/exon fragment argues that this sequence has an important function. One possible function would be to enable splicing under heat shock conditions. If the sequence is necessary for such splicing, one might expect the same conservation in the intron/exon region of the hsp83 transcript, the other heat shock RNA that is known to be spliced. That comparison gives no evidence that the hsrω conservation is necessary for heat shock splicing. The hsp83 sequence has been determined for D. melanogaster, D. pseudoobscura, and D. virilis (5). The hsp83 transcript shows very little sequence conservation in this region. There are only 9/39 matches among the three species at the 3′end of the hsp83 intron. There is a higher level of conservation in the exon (18/first 20 bp of the exon). However the exon is part of the protein coding sequence in hsp83 and the conservation may be explained by the demands of the protein sequence. The second exon of hsrω is not part of an ORF and thus seems likely to be conserved for some other reason (2; 6). One possibility that we are currently testing is that the hsrω intron-exon sequence might serve to regulate splicing and the entry of ω3 into the cytoplasm.

Although the intron-exon segment has the only sequence conservation that is easily detected by cross-hybridization, a comparison of the DNA sequences shows other, much shorter regions of homology between 93D and 2-48B. Hultmark et al., (7) have noted that the D. melanogaster hsp loci, with the exception of hsp 83, have identical nucleotides in the -1, +1, +7, +12, +15, and +20 positions. The three hsrω loci also have the same nucleotides in the first five of these sites, but have a T rather than an A in position +20. The hsrω loci all have 14 nucleotides conserved around the 5′ splice site and 14 nucleotides conserved at the polyadenylation site. All of the regions conserved between D. hydei and D. melanogaster are also conserved in the D. pseudoobscura transcript. These conserved regions suggest that it is important to the cell to be able to make and splice an RNA of this size and to be able to produce it under heat shock conditions.

The ω1 transcripts contain all of the sequence in the ω2 transcript, followed by some "spacer" sequence and several kilobases of a short tandem repeat. Within each Drosophila species the repeats show less than 10% divergence from each other (8; 2). In spite of the homogeneity within a species, the repeat segments of the ω1 transcript are highly divergent, both in the length of the repeat and in its sequence. The D. melanogaster repeat is about 280 bp. The D. hydei repeat is 115 bp. The only shared sequence is a nine nucleotide segment, ATAGGTAGG. Interestingly, the segment is found twice in the 280 bp repeat and once in the 115 bp repeat so that it is distributed about equally along the two RNAs. It seems very likely that this sequence is a binding site for something. If so, the two transcripts may be functionally equivalent in spite of the overall sequence divergence.

RELATIVE LEVELS OF THE hsrω TRANSCRIPTS VARY WITH THE PHYSIOLOGICAL STATE OF THE CELL

One phenotype of the hsrω puff is its induction by agents that do not induce other members of the heat shock puff set. For instance, both benzamide and colchicine induce puffing specifically at hsrω (1). In addition, during the developmental sequence of puffing, there is puffing at 93D. The evidence that hsrω responds to so many agents suggests that it is especially sensitive to environmental conditions. In addition, Lakhotia (1) found that treatments inducing a puff at 93D block subsequent induction by a second agent, if the two inducers are applied in a relatively short time. This evidence that the locus is refractory to a second closely-spaced induction suggests that it is autoregulated.

Analysis of RNA from different developmental stages shows clear evidence of developmental regulation of hsrω (Figure 1 D-E). No hsrω transcripts are detected in very early embryos but they appear to be among the first transcripts of the zygote genome. Levels of hsrω RNA are low in the early larval instars but rise in the third instar and remain high throughout pupal life. The developmental profile of hsrω transcript abundance is reminiscent of ecdysone-induced transcripts, suggesting that the hsrω locus is also sensitive to ecdysone. A D. melanogaster cell line, Schneider line S-3, that is sensitive to ecdysone was used to test ecdysone responsiveness. Ecdysone treatment (10^{-6} M) of these cells induces accumulation of the hsrω transcripts. This induction is not an indirect result of a heat shock since we can detect no induction of hsp 70 RNA. Ecdysone has been shown to induce hsp22 both in S-3 cells and in the intact animal (9). In ecdysone-treated S-3 cells, the rise in hsrω continues for at least 24 hrs, whereas hsp22 RNA increases for 4 hrs and then declines. This suggests that hsrω is involved in a later stage of the ecdysone-induced developmental program.

The slower induction of hsrω by ecdysone also suggests that hsrω transcription is a secondary response to the hormone. It is customary to test for secondary responses by giving ecdysone in the presence of cycloheximide to inhibit synthesis of any new proteins that might be induced by the hormone. We have found that cycloheximide alone leads to high levels of the ω3 transcript. The effect of cycloheximide on ω3 levels

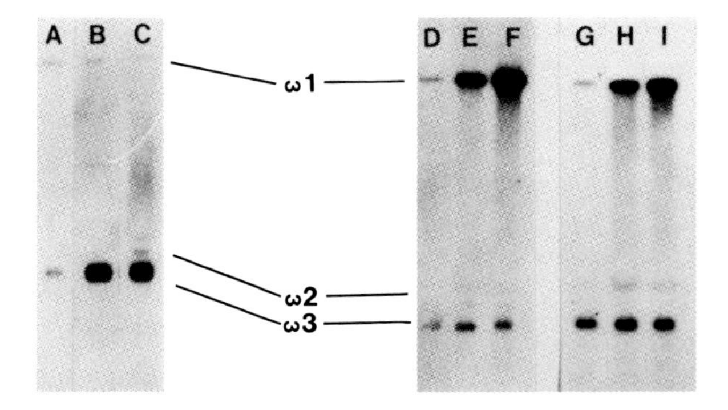

Figure 3. Inhibition of protein synthesis leads to increases in the levels of the hsrω3 transcript; treatment with benzamide or colchicine hyperinduces ω1.. RNA from Schneider S-3 cells was probed with the ω3 sequence. A, control; B and C, cells treated for 2 hours with cycloheximide (10^{-4} M) or pactamycin (10^{-7} M); D and G, control cells; E and F, benzamide (10 mM) for 12 and 24 hrs; H and J, colchicine (100 μg/ml) for 12 and 24 hr.

appears to be due to inhibition of protein synthesis since inhibitors of both initiation (pactamycin) and elongation (emetine, cycloheximide) produce the same result (Figure 3, A-C). The drugs appear to act largely, if not entirely, by inhibiting the turnover of the ω3 transcript. At least during recovery from heat shock, hsrω transcripts turn over much more rapidly than do other RNAs. Drugs affecting translation seem to specifically stabilize the ω3 transcript; we see no significant increase in either the ω2 or the ω1 RNAs. Cycloheximide-induced increases in ω3 levels are seen in salivary glands, as well as in diploid cells, yet we can see no puffing in the 93D region. This lack of puff induction supports the suggestion that cycloheximide is acting only to inhibit turnover of the cytoplasmic transcript.

We have also studied two of the drugs that induce puffing specifically at 93D, colchicine and benzamide. These drugs affect the levels of hsrω transcripts very differently than do the translational inhibitors. Both colchicine and benzamide treatments lead to high levels of the ω1 RNA but produce less dramatic changes in the levels of ω2 and ω3 (Figure 3, D-I). We do not yet know what common feature of these two drugs leads them to affect hsrω in similar ways; however this increase in the ω1 transcript contrasts with the high levels of ω3 induced by translational inhibitors. The different patterns of hsrω transcripts that are induced by different drugs suggests that the relative levels of hsrω transcripts may reflect different disturbances in cellular metabolism.

THE SMALL OPEN READING FRAME IN THE ω3 TRANSCRIPT APPEARS TO BE TRANSLATED.

The ω3 transcript is both spliced and polyadenylated, two characteristics that are generally associated with mRNA, yet we have been unable to detect a polypeptide that might be encoded by this RNA (Figure 4, A-C). The potential open reading frames (ORFs) in the ω3 transcript are all very short in comparison to the size of the transcript. In addition, there is almost no conservation of the ORFs that are seen. The single exception is a small ORF (ORF-ω) beginning near position +120 from the transcription start site. This is the first ORF that is in what is thought to be an appropriate context for translation (10). The first four amino acids in ORF-ω are conserved in all three Drosophila species. The conservation is at the level of amino acids rather than nucleotide sequence. Surprisingly there is little obvious conservation of the rest of the ORF-ω polypeptides; however their lengths are almost the same. ORF-ω in D. melanogaster encodes 27 amino acids, 23 in D. hydei, and 24 amino acids in D. pseudoobscura.

Although we have not detected a polypeptide product of hsrω, we did find that the ω3 transcript sedimented with the monosome/disome peak on polysome gradients. This suggestion that the ω3 transcript was translated was supported by other types of analysis; buoyant density in metrizamide polysome gradients, EDTA release from polysomes, and studies with translational inhibitors. The several kinds of evidence that ω3 was associated with small polysomes, plus the evidence that any disruption of protein synthesis leads to stabilization of ω3 RNA, led us to suspect that ORF-ω was translated (Fini et al., in preparation).

If ORF-ω is translated, it would be expected to inhibit the translation of other ORFs on the same RNA

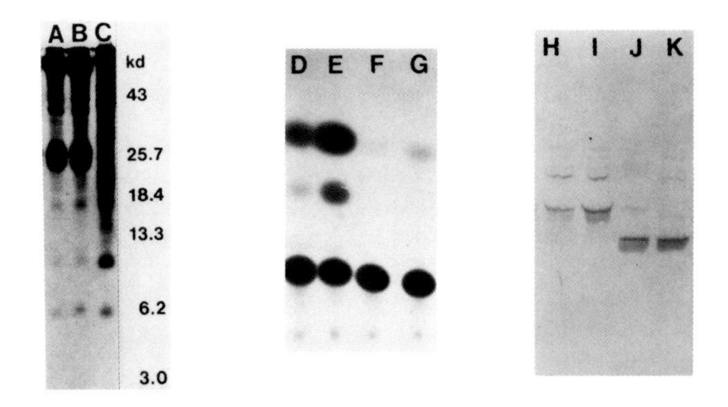

Figure 4. ORF-ω on the hsrω3 transcript appears to be translated although a heat shock induced polypeptide of the expected size is not detected. Proteins from Schneider S-3 cells labeled with ^{3}H-amino acids between 30 and 60 minutes of heat shock (A and B) are compared to control cells (C). Other heat shock proteins are clearly detected. The positions of molecular weight standards are indicated. The leader of D. melanogaster hsrω3 to +67 does not inhibit the translation of a downstream CAT gene at 25°C (D) or at 36°C (E). The D. melanogaster hsrω3 leader to +200 containing ORF-ω does inhibit the translation of a downstream CAT gene at both 25°C (F) and 36°C (G). The D. hydei ORF-ω fused in frame to CAT results in the synthesis of a fusion protein (H and I) larger than the original CAT protein (J and K). Both CAT and the ORF-ω/CAT fusion protein are detected with anti-CAT antibody.

molecule. To test this prediction, two recombinant DNA molecules were constructed utilizing the CAT/SV40 chimeric gene. Both of these constructs contained the upstream control regions of 93D, to ensure that the gene could be induced by heat shock, and part of the ω3 sequence. In the ω-leader-CAT construct, the ω3 sequence ends just in front of ORF-ω and is replaced by the bacterial chloramphenicol acetyl transferase (CAT) gene. In the ORF-ω-CAT construct, the CAT gene had been added 30 nucleotides behind ORF-ω. In both constructs the CAT gene has its own translation start. Both constructs were stably transformed into cultured Drosophila cells.

Although ORF-ω-CAT produced significantly more cytoplasmic RNA than ω-leader-CAT, only the ω-leader-CAT cells made significant amounts of the CAT enzyme (Figure 4, D-G). This result suggested that ORF-ω was being translated and inhibiting translation of the CAT gene immediately behind it. Direct evidence for the translation of ORF-ω came from a third construct in which the CAT coding region was linked in frame to ORF-ω of D. hydei. Cells stably transformed with this construct made a CAT protein that was larger than the wild type protein by the amount expected if translation had started in ORF-ω (Figure 4, H-K).

The ORF-ω-CAT fusion protein is easily detected in the transfected cells. Our hypothesis is that the normal translation product of ORF-ω is not detected because it is turned over very rapidly. Because only the first four amino acids are strongly conserved, it may be that it is the act of translation of ORF-ω, rather than its product that is important to the cell.

THE LOSS OF hsrω APPEARS TO BE DETRIMENTAL TO THE ORGANISM

An attempt to saturate the 93D region with lethal and visible mutations (11) produced no detectable point mutations in the sequences that are now identified as hsrω. This lack of mutations seems less surprising now that we know the hsrω sequence and the way in which it is evolving. It appears that there are only a few regions where nucleotide substitutions would disrupt the functions of the transcripts.

During the mutational analysis, several large deletions that removed hsrω were characterized. Two deletions were identified that seem to be particularly useful. One, Df(3)e^{Gp4}, removes hsrω and 11 complementation groups proximal to it. The other, Df(3)GC14, removes hsrω and 3 complementation groups distal to it. Animals heterozygous for these two deficiencies should be totally lacking only the hsrω locus. Such heterozygotes should make up one quarter of the progeny of a cross between two flies, each carrying one of the deficiencies over a balancer chromosome with an intact hsrω locus. Embryos lacking hsrω hatch as well as siblings of other genotypes; however, the egg was produced by a mother with an intact hsrω locus so we cannot draw conclusions about the need for hsrω during embryogenesis. The lack of hsrω becomes apparent as soon as larvae hatch. Larvae heterozygous for the two deficiencies grow very slowly and most die before pupation. Less than 5% emerge as adults and these die very soon. We are in the process of using P-element-mediated transformation in an attempt to rescue the larvae with the hsrω sequence. If such rescue proves that the phenotype we see is actually due to the loss of hsrω, it will provide clear evidence that hsrω has an important effect on viability.

SUMMARY

In spite of sequence divergence, the hsrω loci have some conserved regions that give clues to their function. These clues suggest that the ω1 transcript could serve as a binding site for a nuclear protein while the ω3 transcript could give a measure of translational activity. It seems reasonable to suggest that hsrω may coordinate nuclear and cytoplasmic activities. Such a role is consistent with evidence that hsrω is very sensitive to environmental conditions and that its loss affects viability.

REFERENCES

1. Lakhotia SC (1987). The 93D heat shock locus in Drosophila: a review. J Genet 66:139.
2. Garbe JC, Bendena WG, Alfano MA, Pardue ML (1986). A Drosophila heat shock locus with a rapidly diverging sequence but a conserved structure. J Biol Chem 261: 16889.
3. Ryseck R-P, Walldorf U, Hoffman T, Hovemann B (1987). Heat shock loci 93D of Drosophila melanogaster and 48B of Drosophila hydei exhibit a common structural and transcriptional pattern. Nuc Acids Res 15:3317.
4. Lindquist S (1986). The heat-shock response. Annu Rev Biochem 55:1151.
5. Blackman RK, Meselson M (1986). Interspecific nucleotide comparisons used to identify regulatory and structural features of the Drosophila hsp82 gene. J Mol Biol 188:499.

6. Garbe JC (1988). Characterization of heat shock locus 93D in Drosophila melanogaster. PhD thesis Massachusetts Institute of Technology, Cambridge, MA
7. Hultmark D, Klemenz R, Gehring WJ (1986). Translational and transcriptional control elements in the untranslated leader of the heat shock gene hsp22. Cell 44:429.
8. Peters FPAMN, Lubsen NH, Walldorf U, Moormann RJM, Hovemann B (1984). The unusual structure of heat shock locus 2-48B in Drosophila hydei. Mol Gen Genet 197:392.
9. Ireland RC, Berger E, Sirotkin K, Yund MA, Osterbur D, Fristrom J (1982). Ecdysterone induces the transcription of four small heat shock genes in Drosophila S-3 cells and imaginal discs. Develop Biol 93:498.
10. Kozak M (1986). Point mutations define a sequence flanking the AUG initiator codon that modulates translation by eukaryotic ribosomes. Cell 44:283.
11. Mohler J, Pardue ML (1984). Mutational analysis of the region surrounding the 93D heat shock locus of Drosophila melanogaster. Genetics 106:249.

Stress-Induced Proteins, pages 15–24

THE CHROMATIN STRUCTURE OF *hsp26*[1]

T.J. Dietz, I.L. Cartwright[2], D.S. Gilmour
E. Siegfried, G.H. Thomas, and S.C.R. Elgin

Washington University
Department of Biology
St. Louis, MO 63130

ABSTRACT The chromatin of the *Drosophila melanogaster hsp 26* gene exists in an ordered structure both before and after heat shock induction. Two DNase I hypersensitive sites (DH sites) are present 5′ to the gene; numbering out from the gene, heat shock elements (HSE) 1,2,6 and 7 lie within these open regions. An ordered nucleosome array is seen downstream in the body of the gene. Base pair resolution mapping of the 5′ region demonstrates the presence of the TATA box binding protein both before and after heat shock. Correspondingly, an exonuclease III protection assay has demonstrated the presence of a general TATA-box binding factor in nuclear extracts from control and heat shocked embyros. Between the two DH sites, a large region of protection (150bp) characterized by 10-11bp periodic DNase I cleavage sites is observed, suggesting the presence of a nucleosome. Following heat shock activation, the transcribed region becomes accessible to DNase I, and the regular nucleosome pattern becomes indistinct (1). The "unfolding" of the chromatin fiber during transcription may be facilitated by the presence of high levels of a topoisomerase I shown to be present during transcription (4). Additional footprints detected over HSE′s 1,2 and 6 reflect the binding of

[1]This work was supported by research grants from the National Institutes of Health (GM 30273), NIHBRSG (507RR07054-22) and the National Science Foundation (DCB-8601449) to S.C.R. Elgin and by a NIH Postdoctoral Fellowship (F32 GM107982) to D.S. Gilmour.
[2]Present address: Department of Biochemistry and Molecular Biology, University of Cincinnati College of Medicine, Cincinnati, Ohio 45267.

the heat shock transcription factor. This factor is detected in significant amounts only in nuclear extracts from heat shocked embryos. We suggest that the binding of the TATA-box factor and of a specifically positioned nucleosome generates the upstream DNase I hypersensitive sites, leaving the HSE's accessible; the folding of the DNA by the nucleosome brings these sites into physical proximity, perhaps facilitating gene activation.

INTRODUCTION

Various types of stress have been shown to induce a family of heat shock genes in *Drosophila*. These genes offer a simple model system for the study of inducible gene expression. The four small heat shock genes, *hsp22*, *23*, *26* and *28*, in addition to being activated by environmental stresses are activated in response to a set of predetermined developmental signals in the fly. The 5′ regulatory region of one of these, *hsp26*, encompasses several common promoter features required and some unique elements. We have focussed our study on this gene, analyzing the protein-DNA interactions *in vivo* and *in vitro* to generate a model of the "activatable" gene.

RESULTS AND DISCUSSION

Chromatin Structure of *hsp26*

Insight into the mechanisms governing eucaryotic gene regulation can be gained by determining the specific protein-DNA interactions at a gene before and after transcriptional activation. The structural changes occurring in the chromatin during heat shock induction of the *Drosophila hsp26* gene have been analyzed by the use of two DNA cleavage reagents, DNase I and methidiumpropyl EDTA·Fe(II) [MPE·Fe(II)] (1). The accessibility to cleavage reagents of the DNA identifies those regions that are relatively free of protein. Nuclei are isolated from non-heat shocked and heat shocked *Drosophila* embryos and a brief digestion of the nuclear chromatin is performed. The locations of DNA cleavage sites in nuclear chromatin are then determined using the technique of indirect end-labelling (as described in the legend to Figure 1)(1). The positions of the DNA cleavage sites in nuclear chromatin for

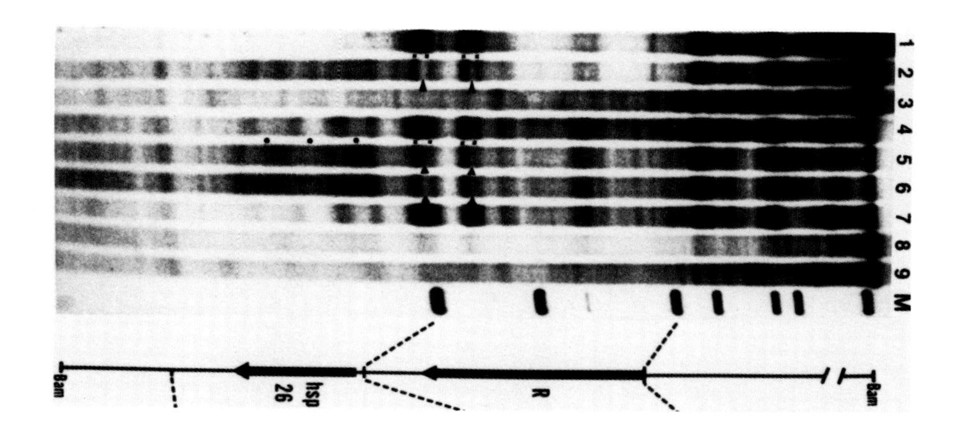

Figure 1. Chromatin mapping in the vicinity of the hsp26 gene. Nuclei were isolated from non-heat shocked (NHS) and heat shocked (HS) *Drosophila* embryos. DNA digestions were at 25°C with DNase I (lanes 1-3) or MPE·Fe(II) (lanes 4-9). Isolated DNA was digested with BamHI. DNA samples (9μg) were fractionated on a 1.2% agarose gel; the gel blotted to nitrocellulose and the filter hybridized to a probe abutting the BamHI site downstream of hsp26. DNase I digestions of NHS nuclei (lane 1), HS nuclei (lane 2), and genomic DNA (lane 3); MPE·FE(II) digestions of NHS nuclei (lane 4), HS nuclei (lane 5-6) 0.35M KCl-extracted NHS nuclei (lane 7), 0.5M KCl-extracted NHS nuclei (lane 8), and genomic DNA (lane 9). M denotes molecular weight markers. Squares denote hypersensitive sites, circles mark nucleosome positioning and arrowheads denote the heat shock induced footprint in the proximal and distal hypersensitive sites. Adapted from ref. 1.

hsp26 are shown in Figure 1. Mapping of the chromatin of the hsp26 gene reveals an ordered structure both prior to and subsequent to heat shock induction. Two DNase I hypersensitive sites reside 5′ to the gene; heat shock elements 1,2,6 and 7 (numbering out from the gene) lie within these open regions. The body of the gene appears to be associated with an ordered array of nucleosomes. This can only be detected by MPE·Fe(II); DNase I does not appear to have access to the DNA in this region.

Upon gene activation parts of the DH sites become protected against cleavage, indicating the presence of a protein protected DNA region (arrowheads in Figure 1). Results using MPE·Fe(II) are very similar. The protection can be attributed to binding of the heat shock factor (HSF) to the HSE′s (2,3). In addition, the ordered array of nucleosomes appears to be disrupted since the region becomes more accessible to DNase I and MPE·Fe(II). This change is gene-specific; neither access for DNase I nor smearing of the MPE·Fe(II) pattern is seen in the transcription unit of the adjacent developmentally regulated R gene, which is not induced at high levels upon heat shock. The perturbations detected in the nucleosome structure of the chromatin appear to be concomitant with transcriptional activation (1). The physical extent of these alterations coincides with a region associated with topoisomerase I, which interacts with both strands of the activated gene at high levels (4).

Genomic Footprinting of hsp26

The high resolution sequence gel blotting technique of Church and Gilbert (5) has been used to determine the protein-DNA interactions occurring in the chromatin fiber of hsp26 at the sequence level (Figure 2) (6). A TATA box binding protein is detected in nuclei isolated from both control embryos and heat shocked embryos. Interestingly, the footprint covers only one half of the TATA box and extends slightly downstream. We have recently observed a TATA box binding activity in ammonium sulfate extracts from embryo nuclei that has a similar 3′ boundary as that detected in vivo (7). In contrast to the TATA box binding activity, occupation of the HSE′s, presumably by the HSF, is observed only upon heat shock induction. Only HSE′s 1/2 and 6 are observed to be occupied upon heat shock; previous results from deletion analysis show these sites are essential for heat shock induction. Sites 3,4,and 5 contribute little if any to the heat inducibility of hsp26 (8) and are observed here to be unoccupied by HSF. Binding

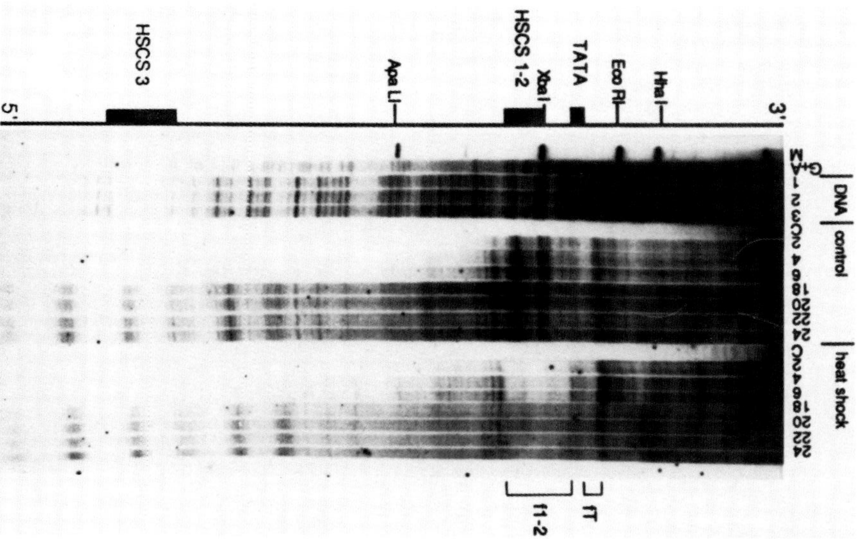

Figure 2. Genomic footprinting of the proximal region of the hsp26 promoter. Purified nuclear DNA from DNase I digested chromatin is cut 5′to the heat shock consensus sequence (HSCS) 3. A probe abutting this distal site is used to visualize sequences from ca. -170 to within the leader sequences. Cleavage in the chromatin is on the sense strand. 'Control' lanes denote nuclei isolated from non-heat shocked embryos. 'Heat shock' denote nuclei isolated from heat shocked embryos. DNA in nuclei were digested with DNase I using 4-24 U/ml as indicated above the lanes. C-no DNaseI control; DNA-purified DNA digested with DNase I; C+T and G+A are genomic sequencing markers; M-markers generated by digestion of genomic DNA with the indicated restriction enzyme plus the enzyme used to make the distal 5′ cut. Brackets labelled fT and f1-2 are footprints encompassing the TATA box and HSCS 1 and 2 respectively. fT is constitutive, while f1-2 is heat shock specific. Adapted from ref. 6.

of the HSF appears to have influenced the interaction of the TATA box binding protein with the DNA, as 3′ to the TATA box footprint a series of enhanced DNase I cleavage events is observed on both strands. The HSF might act by interacting with the TATA box binding protein and causing the effective release of RNA polymerase. An RNA polymerase molecule has been shown to be poised on both the hsp70 and hsp26 promoters prior to induction (9) (A. Rugvie and J.T. Lis, pers. comm.).

Approximately 200bp of DNA lies between the HSE′s critical for gene activation. This region is relatively resistant to DNase I digestion; however, with high levels of digestion a very characteristic 10-11bp periodic cleavage pattern is detected, which indicates that this region of the DNA is bound to a surface. When nuclear chromatin is digested with micrococcal nuclease, the DNA between the two DH sites is recovered as a fragment 150-160bp in size. Together, the DNase I and micrococcal nuclease digestion patterns suggest that the region of DNA located between the essential HSE′s is bound to a precisely positioned nucleosome. Either a bound protein or the DNA structure itself might provide the signal for the precise positioning of the nucleosome. Proximal to the putative nucleosome is a homopurine-homopyrimidine stretch that is known to exist in a non-B form DNA structure *in vitro* under appropriate conditions of supercoiling and pH (10). If a non-B form configuration were adopted *in vivo*, making the DNA unfit for incorporation into a nucleosome, it would then create a boundary at which the nucleosome could be positioned.

In vitro Characterization of *hsp26* Regulatory Components

The identification and characterization of the proteins responsible for the chromatin structure of the regulatory region will be important for a complete understanding of the process of *hsp26* transcriptional activation. Using nuclear extracts from both heat shocked and non-heat shocked embryos, we have monitored protein-DNA interactions on the *hsp26*, *hsp70*, histone *H3* and histone *H4* promoters (7). Both extracts transcribe each of these promoters at comparable levels. An exonuclease III protection assay (11) has been used to detect specific protein-DNA interactions on these promoters; a DNA fragment that has been radiolabelled at one end is incubated with the nuclear extract and then digested with exonuclease III. Tightly bound proteins will prevent the exonuclease from digesting the DNA fragment. The location of these exonuclease "barriers" as determined by

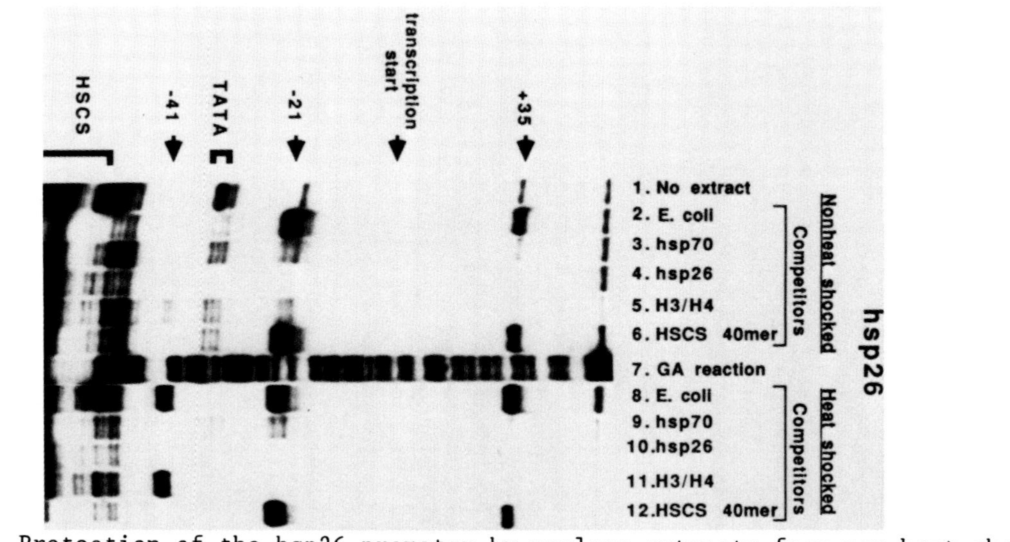

Figure 3. Protection of the hsp26 promoter by nuclear extracts from non-heat shocked (NHS) and heat shocked (HS) embryos from exonuclease digestion. A radiolabelled hsp26 promoter fragment was incubated (25°C for 10′) with the nuclear extracts from NHS (lanes 2-6) or HS (lanes 8-12) embryos. Additionally, either 200ng of HaeIII cut E. coli DNA (lanes 2 and 8), hsp70 (lanes 3 and 9), hsp26 (lanes 4 and 10), H3/H4 (lanes 5 and 11) or a plasmid containing 40 copies of hsp70's heat shock consensus sequences (HSCS 40mer; lanes 6 and 12) were present as competitor. Subsequent to incubation all samples were treated with exonuclease III (30°C for 10′). DNA was purified and analyzed on a sequencing gel. Lane 1 contains DNA digested with exonuclease but with no extract added. Lane 7 is a purine cleavage reaction done on the end-labelled DNA. The transcription start site is labelled. TATA - TATA box homology; HSCS - heat shock consensus sequence. The numbered arrows (numbered relative to the transcription start) denote positions where protein-DNA interactions impede the digestion by exonuclease.

sizing the fragments on a sequencing gel indicates the positions of bound protein. A striking difference is observed between the barriers that occur on purified hsp26 promoter DNA and those that occur when nuclear extract is present (see Figure 3). Two prominent barriers are present at positions +35 and -21 (relative to the start of transcription) when using extracts from non-heat shocked embryos, while an additional barrier at -41 is observed with extracts from heat shocked embryos. The barrier at -21 flanks the TATA box homology, while the barrier at -41 flanks the HSE. The +35 and -21 barriers are also observed to occur, in very similiar positions, in the hsp70, histone H3 and histone H4 promoters. The barrier bordering the HSE in hsp26 seen using extracts from heat shocked embryos is also detected on hsp70, flanking its HSE in an analogous fashion.

Competition reactions with nonradioactive DNA fragments indicate that the barriers produced on hsp26 in the presence of nuclear extract are due to sequence-specific DNA binding proteins. When excess promoter fragments from hsp26, hsp70, histone H3 or histone H4 are added, the barriers at +35 and -21 are significantly reduced, while excess E. coli DNA or a fragment containing 40 copies of the hsp70 HSE's (HSCS 40mer) have no effect. In contrast, the barrier at -41, adjacent to the HSE, is diminished by competition with nonradioactive excess hsp26 or hsp70 promoter fragments and by the HSCS 40mer, and not with E. coli, histone H3 and histone H4 competitor DNA's. It can be concluded that the barrier at -41 is due to binding of a protein to the HSE.

The TATA box, which is required for in vitro transcription, is also required for the binding of factors which cause the barriers at +35 and -21. When the TATA box of the hsp70 promoter is deleted, not only does the barrier at -21 disappear but the +35 barrier as well is absent. This deleted promoter fragment is not transcribed in vitro. When sequences downstream of +7 in hsp26 are replaced with sequences from pUC13, the barrier at +35 still forms. Thus there appears to be no sequence requirements downstream of +7 for the barrier to +35 to form. This promoter is transcribed at levels comparable to that of the normal hsp26 promoter.

Our work with the nuclear extracts suggests that two binding activities, represented by the exonuclease barriers at -21 and +35, require the TATA box homology. These elements are present in extracts from both non-heat shocked and heat shocked embryos, indicating that they might function prior to trasncriptional activation. Moreover, these binding activities may be due to general transcription

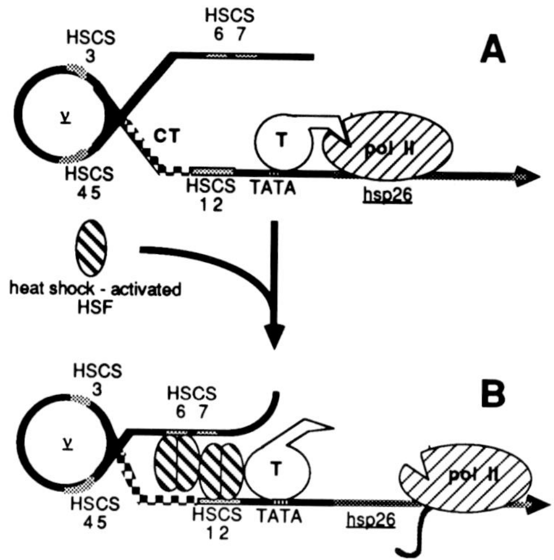

Figure 4. Model for hsp26 transcriptional activation. Panel A-hsp26 promoter prior to heat shock; Panel B-hsp26 promoter subsequent to heat shock. v-nucleosome, T-TATA box binding factor, pol II-RNA polymerase II molecule (with RNA in B), HSF-heat shock factor; HSCS-heat shock consensus sequence, TATA-TATA box homology; CT-homopurine-homopyrimidine stretch. Adapted from ref. 6.

factors, since binding also occurs on the hsp70, histone H3 and histone H4 promoters. The heat shock factor detected at high levels in extracts from heat shocked embryos probably plays a critical role in the transcriptional activation.

A Model of hsp26 Induction

Figure 4 presents a model for protein-DNA architecture that compares the hsp26 promoter region before and after transcriptional induction. One RNA polymerase II molecule appears to be present at the hsp26 and hsp70 promoters prior to heat shock (9) (A. Rugvie and J.T. Lis. pers. comm.) As shown by our work reviewed here we observe a TATA box binding activity before and after heat induction, and an HSE binding activity after heat shock. The region of the DNA between HSE's 1/2 and 6 gives indications of an exactly positioned nucleosome. This nucleosome and the binding of the TATA box factor could generate the proximal DNase I hypersenstive site, thereby leaving the HSE's accessible. Upon heat shock, HSF binds to the accessible HSE's. We propose that the winding of the DNA around the nucleosome brings into juxtaposition site 6 with site 1/2 and thus places both sites in close proximity with the poised polymerase. Subsequent binding of closely juxtaposed HSF's provides a releasing mechanism by which polymerase can now engage in transcription.

REFERENCES

1. Cartwright IL, Elgin SCR (1986). Nucleosomal instability and induction of new upstream protein-DNA associations accompany activation of four small heat shock protein genes in Drosophila melanogaster. Mol Cell Biol 6:779.
2. Wu C, Wilson S, Walker B, Dawid I, Paisley T, Zamarino V, Veda H (1987). Purification and properties of Drosophila heat shock activator protein. Science 238:1247.
3. Parker CS, Topol J (1984). A Drosophila RNA polymerase II transcription factor binds to the regulatory site of an hsp70 gene. Cell 36:273.
4. Gilmour DS, Elgin SCR (1987). Localization of specific topoisomerase I interactions within the transcribed region of active heat shock genes by using the inhibitor camptothecin. Mol Cell Biol 7:141.
5. Church GM, Gilbert W (1984). Genomic sequencing. Proc Natl Acad Sci USA 81:1991.
6. Thomas GH, Elgin SCR (1988). Protein/DNA architecture of the DNase I hypersensitive region of the Drosophila hsp26 promoter. EMBO J 7: 2191.
7. Gilmour DS, Dietz TJ, Elgin SCR (1988). TATA box-dependent protein-DNA interactions are detected on heat shock and histone gene promoters in nuclear extracts derived from Drosophila embyros. Manuscript submitted.
8. Cohen RS, Meselson M (1985). Separate regulatory elements for the heat-inducible and ovarian expression of the Drosophila hsp26 gene. Cell 43:737.
9. Gilmour DS, Lis JT (1986). RNA polymerase II interacts with the promoter region of the non-induced hsp70 gene in Drosophila melanogaster cells. Mol Cell Biol 6:3984.
10. Siegfried E, Thomas GH, Bond UM, Elgin SCR (1986). Characterization of a supercoil-dependent S1 sensitive site 5' to the Drosophila melanogaster hsp26 gene. Nuc Acids Res 14:9425.
11. Wu C (1985). An exonuclease protection assay reveals heat-shock element and TATA box DNA-binding proteins in crude nuclear extracts. Nature (London) 371:84.

II. TRANSCRIPTIONAL REGULATION
A: PROCARYONS

Stress-Induced Proteins, pages 27–36

DnaK AND GroE PROTEINS PLAY ROLES IN *E. coli* METABOLISM AT LOW AND INTERMEDIATE TEMPERATURES AS WELL AS AT HIGH TEMPERATURES

Bernd Bukau, Caroline E. Donnelly, and Graham C. Walker

Biology Department, Massachusetts Institute of Technology
Cambridge, MA 02139

ABSTRACT

We have analyzed cellular defects caused by mutations in *dnaK* and *groE* heat shock genes and show that DnaK and GroE proteins play roles in *E. coli* metabolism at low and intermediate as well as at high growth temperatures. Δ*dnaK* mutants were cold sensitive as well as temperature sensitive for growth. At intermediate temperature (30°C), introduction of a Δ*dnaK* allele into wild type caused severe defects in cell division, slow growth and poor viability of the cells. Overproduction of FtsZ protein suppressed cell division defects of Δ*dnaK* mutants; however, slow growth and poor viability at 30°C and temperature sensitivity of growth were not suppressed, indicating additional functions are defective in Δ*dnaK* mutants besides cell division. Δ*dnaK* mutants were unstable at 30°C and frequently acquired secondary mutations which were unlinked to *dnaKJ*. These secondary mutations suppressed cell division defects, poor viability and slow growth at 30°C, but not temperature sensitivity of growth of Δ*dnaK* mutants.
Mutations in *groE* were able to suppress cold sensitivity caused by overproduction of the UmuC and UmuD proteins which are required for UV mutagenesis. Furthermore, *groE* mutants were reduced in UV mutagenesis at 30°C, and nearly nonmutable at 42°C, indicating a function of the GroE proteins in UV mutagenesis. Suppression of cold sensitivity of cells overproducing UmuC, D proteins by mutations in *groE* might result from absence of this function.

INTRODUCTION

DnaK, GroEL and GroES are the most abundant of 17 heat shock proteins in *E. coli* (11,12). DnaK is a 69 kd protein that belongs to the ubiquitous Hsp70 family and shares about 50% homology with human and Drosophila Hsp70 proteins (11,12). GroEL and GroES are proteins with molecular weights of 63 kd and 11 kd, respectively. GroEL possesses antigenic crossreactivity to newly identified mitochondrial proteins in a number of organisms including yeast and humans (10). This high conservation of DnaK and GroE proteins during evolution suggests their biological importance.

DnaK and GroE proteins play essential roles in growth of various bacteriophages. In the case of phage λ, DnaK is required for the initiation of replication of λ DNA and appears to act by dissociating DnaB from the λ P protein thereby allowing the helicase to act (4,5). The GroE proteins are required for assembly of the phage head during morphogenesis of λ (5).

The cellular functions of the DnaK and GroE proteins, however, are not understood. After heat shock to 45°C, DnaK and GroE proteins become extremely abundant, accounting for more than 15% of the protein mass of the cell. Both proteins are essential for growth at high temperature, since most *dnaK* and *groE* mutants are conditional and temperature sensitive for growth. A shift of these mutant cells to the nonpermissive temperature causes inhibition of synthesis of DNA and RNA (7,18) and a block of cell division leading to the formation of cell filaments (13,17). Sakakibara (15) has recently presented evidence suggesting that DnaK functions in the initiation of chromosomal replication at least at high temperatures. Furthermore, DnaK plays a role in regulation of the heat shock response since *dnaK* mutants fail to turn off the heat shock response after shift to 42°C (16).

At temperatures below heat shock temperatures the cellular functions of the DnaK and GroE heat shock proteins are even less clear. The fact that they are expressed at considerable levels below 42°C, e.g. constituting 3% of the protein mass of cells grown at 37°C (11), suggests that they are important not only after heat shock. Further evidence comes from analysis of deletion mutants in *htpR* which encodes the activator of expression of the heat shock genes, σ^{32}. Δ*htpR* mutants are largely deficient in synthesis of heat shock proteins and unable to grow at temperatures above 20°C (T. Yura, C. Gross, pers. communication). Their growth temperature range can be extended to up to 42°C by mutations which lead to the overproduction of GroE and DnaK (N. Kusukawa and T. Yura, pers. communication). Thus, the DnaK and GroE proteins seem to possess cellular functions at lower and intermediate temperatures and to be the biologically most important heat shock proteins.

In this paper we will review our recent results which will be published elsewhere demonstrating that the DnaK and GroE proteins play roles in cellular physiology at lower and intermediate temperatures as well as after heat shock.

Δ*dnaK52* MUTANTS ARE COLD SENSITIVE AS WELL AS TEMPERATURE SENSITIVE FOR GROWTH

The *dnaK* gene is organized in an operon together with the promoter-distal *dnaJ* heat shock gene. Our laboratory recently reported the construction of a deletion of the *dnaK* gene (Δ*dnaK52*), which removed the promoter region and 933 base pairs of the coding region of the *dnaK* gene (13). A major conclusion of this work was that phenotypes of Δ*dnaK52* mutants were similar to those of previously isolated conditional *dnaK* mutants in that they were unable to grow at 42°C but could grow at 30°C. However, the experiments summarized here indicate that deleting the *dnaK* gene causes severe cellular defects even at 30°C and promted us to reinvestigate the requirement for the *dnaK*$^+$ function at various growth temperatures.

In attempting to introduce the Δ*dnaK52* allele into wild type cells by P1*vir* transduction at 30°C, we observed that the yields of Δ*dnaK52* transductants were very low. It turned out that the Δ*dnaK52* allele was transduced with particularly poor efficiency, as compared to other chromosomal markers. Furthermore, Δ*dnaK52* transductants grew very slowly, formed flat and translucent colonies, restreaked poorly, and yielded faster growing variants during the purification of the transductant colonies, as described below in more detail. These findings indicate that Δ*dnaK52* mutants have considerable growth defects even at 30°C which may account for the low transduction efficiency of the Δ*dnaK52* allele at that temperature.

We addressed the question of whether it might be possible to find conditions where DnaK is completely dispensable for cellular growth and reasoned that, as in the case of σ^{32}, the cellular requirement for DnaK might decrease with decreasing growth temperature. To systematically investigate the requirement for DnaK at various temperatures, we transduced the Δ*dnaK52* allele into wild type cells at a series of temperatures from 11°C to 42 °C. Surprisingly, we could not cross the Δ*dnaK52* allele into wild type cells at low (11°C, 16°C) as well as at high (42°C) temperatures, unless a second, intact copy of the *dnaKJ* operon was present. Furthermore, fresh Δ*dnaK52* transductants which were isolated at 30°C were unable to grow either at high (42°C) or low (11°C, 16°C) temperatures. Thus, Δ*dnaK52* mutants have a narrow temperature spectrum for growth, from around 21°C to temperatures below 42°C. This suggests that the DnaK protein is required to extent the spectrum of temperatures over which *E. coli* can grow.

INTRODUCTION OF THE Δ*dnaK52* ALLELE INTO WILD TYPE CELLS RESULTS IN CELL DIVISION DEFECTS AT 30 °C

As mentioned above, Δ*dnaK52* transductants isolated at 30°C formed unusually flat and translucent colonies. Surprisingly, the cells contained within these colonies formed very long filaments. Electron microscopic analysis revealed that these filaments were unseptated, indicating that cell division was blocked at an early stage prior to septum formation. DNA was segregated within these filaments, but we cannot exclude the possibility that DNA distribution is unequal. We could show that filamentation of Δ*dnaK52* mutant cells was due to lack of DnaK and not to reduction of expression of DnaJ, which might be expected to result from polar effects of the Δ*dnaK52* mutation on expression of the promoter-distal *dnaJ* gene.

The finding that introduction of the Δ*dnaK52* allele into wild type causes filamentation of the cells even at 30°C indicates that the DnaK protein is required, directly or indirectly, for cell division in general, and not only for cell division occurring after heat shock.

OVERPRODUCTION OF FtsZ PROTEIN SUPPRESSES FILAMENTATION OF Δ*dnaK52* MUTANTS AT 30 °C

We tested three alternative hypotheses to explain the cell division defects of Δ*dnaK52* mutants. First, absence of the DnaK protein causes induction of a cellular response that inhibits cell division. We considered two inducible cellular responses, the stringent response and the SOS response, which both were known to interact with the heat shock response (6,8), and therefore might be likely candidates for causing filamentation of Δ*dnaK52* mutants. However, Δ*dnaK52* transductants of *relA1* and *lexA3* mutants, which were unable to induce the stringent response and the SOS response, respectively, were filamenting, slow growing and restreaked poorly at 30°C, and remained temperature sensitive for growth. While these experiments show that the stringent response and the SOS response are not involved in filamentation of Δ*dnaK52* mutants, we cannot rule out participation of other inducible cellular responses.

As a second hypothesis we considered the possibility that the DnaK protein is normally required for initiation of DNA replication at 30°C and, as a consequence of the coupling of DNA replication with cell division, its absence in Δ*dnaK52* transductants leads to filamentation of the cells. This hypothesis was supported by Sakakibara's finding (15) that a temperature sensitive *dnaK* mutant is blocked in initiation of DNA replication after a shift to the nonpermissive temperature. In an *rnh* background, where chromosomal DNA replication is driven by a DnaA protein- and *oriC*-independent process, a shift to the nonpermissive temperature did not lead to inhibition of DNA replication of the *dnaK* mutant. This result suggests that the DnaK protein plays a role in the regular, DnaA- and *oriC*-dependent initiation of DNA replication, but not in DnaA- and *oriC*-independent initiation processes. However, we found that Δ*dnaK52* transductants of an *rnh* strain filamented, grew slowly and were poorly viable at 30°C, and remained temperature sensitive for growth, demonstrating that these phenotypes were not caused by inhibition of DnaA- and *oriC*-dependent initiation of DNA replication.

As a third hypothesis, we considered the possibility that the DnaK protein is normally involved in cell division, and its absence leads to a defect in division. We specifically asked whether it might be possible to suppress filamentation of Δ*dnaK52* transductants by increasing the frequency of septation. In wild type cells the FtsZ protein is limiting for septation and its overproduction causes extra septations leading to formation of minicells (20). We transduced the Δ*dnaK52* allele at 30°C into wild type cells containing the plasmid pZAQ, which carries the *ftsQAZ* operon and causes overproduction of the FtsZ, A, and Q proteins. Microscopic analysis of the resulting transductants revealed that filamentation was greatly suppressed. Suppression of filamentation was due to overproduction of the FtsZ protein since derivatives of pZAQ which eliminate expression of the *ftsZ* gene were ineffective in suppression. Suppression of *dnaK*-mediated cell division defects by overproduction of a component of the regular cell division machinery indicates that the DnaK protein is normally directly or indirectly involved in cell division. Experiments to determine whether DnaK affects the cellular concentration or activity of the FtsZ protein are underway.

SUPPRESSION OF CELL DIVISION DEFECTS DOES NOT SUPPRESS GROWTH DEFECTS OF Δ*dnaK52* MUTANTS AT 30 °C AND 42 °C

Despite efficient suppression of filamentation of Δ*dnaK52* transductants at 30°C by pZAQ, the colonies grew very slowly and restreaked poorly. Furthermore, prolonged incubation of the plates led to frequent appearance of faster growing papillae within the transductant colonies, indicating that additional mutations can further increase the growth rate of Δ*dnaK52*(pZAQ) cells. Also, the presence of pZAQ did not allow Δ*dnaK52*(pZAQ) transductants to grow at 42°C. These findings demonstrate that defects in cell division are not the only cellular defects of Δ*dnaK52* mutants at 30°C and 42 °C.

SPONTANEOUS SECONDARY MUTATIONS SUPPRESS FILAMENTATION AND GROWTH DEFECTS OF Δ*dnaK52* MUTANT CELLS AT 30°C

As mentioned above, upon further incubation at 30°C of plates containing flat and translucent Δ*dnaK52* transductant colonies, papillae appeared in most colonies. Cells from these papillae restreaked well, grew faster and formed colonies of normal morphology. We were unable to stably maintain the flat, translucent colony phenotype of the original Δ*dnaK52* transductants during purification since better growing variants arose so frequently. Cells isolated from the papillae had a strongly reduced filamentation phenotype. The sizes of cells were usually heterogenous ranging from shorter filaments to minicells. Thus, it appeared that septation events in cells isolated from the papillae are much more frequent than in fresh, filamenting Δ*dnaK52* transductant cells, but that the locations of septa are more random than in wild type cells.

We then determined whether reduction of filamentation and higher viability of cells isolated from papillae of Δ*dnaK52* transductants was due to secondary mutations. We found that eight isolates tested all contained secondary mutations which were not linked to the *dnaKJ* operon. We refer to this class of suppressor mutation as being of the class I type. For further genetic analysis of the class I suppressors, we chose one isolate, BB1130. We mapped the suppressor mutations of strain BB1130 and found that it seemed to contain two mutations, both located in the 90 min. region of the *E. coli* chromosome. Crossing out either suppressor mutation resulted in cells that filamented and formed flat and translucent colonies which restreaked poorly.

These results show that Δ*dnaK52* mutants cannot be stably maintained at 30°C without secondary mutations. Δ*dnaK52* isolates which have been described recently (13) most likely contained one or more class I suppressor mutations.

Δ*dnaK52* MUTANTS CARRYING CLASS I SUPPRESSOR MUTATION(S) ARE COLD SENSITIVE AS WELL AS TEMPERATURE SENSITIVE FOR GROWTH

We then adressed the question whether class I suppressor mutations allow growth of Δ*dnaK52* transductants at low (11°C, 16°C) and high (42°C) temperatures. We found that all Δ*dnaK52* isolates we tested that carried class I suppressors were still temperature sensitive for growth, and most but not all were cold sensitive, similar to Δ*dnaK52* mutants.

By selecting for survivors at 16°C or 42°C of temperature and cold sensitive Δ*dnaK52* isolates carrying class I suppressors we isolated cold resistant or temperature resistant mutants. The mutations which allow growth at high or low temperatures were provisionally named class II suppressors. Some, but not all mutants carrying class II suppressors isolated at 42°C could grow at 16°C. This result indicates that the biological functions of the DnaK protein at high growth temperature are different from those at low growth temperature.

GroEL and GroES PROTEINS ARE INVOLVED IN UV MUTAGENESIS

Exposure of *E. coli* to UV irradiation and other mutagens results in the increase in expression of genes that are members of the SOS regulon (19). This induction is required to repair DNA damage and to resume growth of the cells. One component of this repair process is called "SOS processing" and this process is mutagenic (19). For some time our laboratory has been interested in the molecular basis of SOS processing which requires the *recA*, *lexA* and *umuDC* gene products. While investigating the functions of the *umuDC* gene products we have found that two heat shock proteins (the products of the *groEL* and *groES* genes) appear to affect the activity of the *umuDC* gene products.

The *umuDC* genes are repressed by the LexA protein and are overexpressed when present on a high copy number plasmid in a *lexA51*(Def) *sulA* background. Such a strain grows well at 42°C, but when the culture is shifted to 30°C DNA synthesis stops and the cells filament (9). We suspect that the *umuDC* products, when overproduced, interfere with DNA synthesis at the replication fork and prevent the continuation of replication at 30°C. Mutations in several heat shock genes (*htpR*, *dnaK*, *lon*, *grpE*, *groEL* and *groES*) suppress this cold sensitive phenotype.

We suspected that proteolysis may be involved in the suppression of cold sensitivity due to overexpression of *umuDC*, since mutations in many heat shock genes lead to alterations in cellular proteolysis (1). We also have some evidence that the LexA51(Def) protein is stabilized in a *lon*::Tn*10* background (C. Dykstra, pers. communication). If the half-life of the LexA51(Def) protein were lengthened in these heat shock mutants, expression of *umuDC* would be repressed and thus lead to suppression of cold sensitivity. We compared the suppression of cold sensitivity in a *lexA51*(Def) background and in a *lexA*::Tn*5* background and found that most of the heat shock mutants were not able to suppress cold sensitivity in the *lexA*::Tn*5* strain. When transcription of the *umuDC* operon is under the control of the IPTG inducible Ptac promoter we found the same result. Mutations in *dnaK*, *grpE* and *lon* appear to stabilize the LexA51(Def) protein and repress expression of *umuDC*. In contrast, mutations in *htpR*, *groEL* and *groES* suppress cold sensitivity regardless of the *lexA* allele or the transcription regulation.

We have become quite interested in the role that the *groE* genes play in phenotypes associated with the *umuDC* genes. The *groEL* and *groES* mutations may affect the activity (either directly or indirectly) of the UmuD and UmuC proteins or alter the expression of the genes or the stability of the proteins. In examining the activity of the *umuDC* genes in the *groE* background we have found that mutability in the groE strains is reduced more and more as the temperature is increased. At 42°C the *groE* strains are virtually nonmutable by UV light. Secondly, these experiments showed that the *groE* strains also became markedly more UV sensitive than the wild type as the temperature was increased. The defect in UV mutability and UV sensitivity due to the *groE* mutations can be alleviated in the presence of multicopy *umuDC*, so we feel that the *groE* gene products may effect UmuD and/or UmuC directly.

DISCUSSION

In this study we have described various cellular defects caused by mutations in *dnaK* and *groE*. Δ*dnaK52* mutants were cold sensitive as well as temperature sensitive for growth. This demonstrates essential cellular functions of DnaK protein for growth at the extreme ends of the growth temperature range of *E. coli*, underlining its role as a temperature stress protein. It is not clear, however, what the metabolic processes are that require DnaK at high or low temperatures. The requirements for DnaK at high temperature seem to be different from those at low temperature, since secondary mutations which suppress temperature sensitivity of growth of Δ*dnaK52* mutants do not always suppress cold sensitivity of these cells.

Cold sensitivity of Δ*dnaK52* mutants is puzzling in view of the recent finding that Δ*htpR* mutants, which lack the alternative sigma subunit (σ^{32}) required for the heat shock response were able to grow only at temperatures below 20°C. This apparent paradox might be explained by assuming that the residual synthesis of DnaK which still is occurring in Δ*htpR* mutants (C. Gross, pers. communication) is sufficient for growth of Δ*htpR* mutants at temperatures below 20°C.

In addition to essential cellular functions at low and high growth temperatures, DnaK plays important roles in cellular metabolism at intermediate temperatures. This is indicated by the facts that Δ*dnaK52* transductants grow slowly, have a low viability and are genetically very unstable at 30°C. The most striking defect of Δ*dnaK52* mutants was inhibition of cell division. This defect seems to be at or prior to septation since overproduction of FtsZ protein, which normally is required for septation, can suppress the division defects. While this suppression might be explained in several ways, one of the most economical hypotheses is that DnaK directly or indirectly affects the expression or function of FtsZ.

In addition to defects in cell division, other cellular functions must be defective in Δ*dnaK52* mutants at 30°C. This can be concluded from the fact that suppression of cell division defects by overproduction of FtsZ does not suppress slow growth and low viability at 30°C of Δ*dnaK52* mutants. These additional defects seem to be severe, since spontaneous suppressor mutations occur frequently, even when cell division defects are suppressed.

Analysis of spontaneous secondary mutations present in genetically stable Δ*dnaK52* isolates showed that probably two suppressor mutations were present in one isolate. While presence of these mutations allowed better growth and suppression of cell division defects of Δ*dnaK52* mutants at 30°C, growth of these cells at 42°C or 11°C and 16°C required additional mutations. The requirement for these additional mutations for growth at high and low temperatures again indicates various cellular functions of DnaK.

The GroE proteins appear to be involved at some level in the cellular process of UV mutagenesis suggesting that, like DnaK, these proteins play pleiotropic roles in normal cellular metabolism. Like certain of the DnaK functions discussed above, this requirement for GroE in mutagenesis does not seem to be absolute since it can be suppressed by overexpression of the UmuDC proteins.

We do not yet understand the role of the DnaK and GroE proteins in these various cellular processes. However, it is possible that they act by mediating certain intermolecular protein-protein interactions (i.e. by participating in the association or dissociation of particular proteins) or certain intramolecular polypeptide interactions (i.e. by participating in certain protein folding events), similar to the roles that DnaK plays in initiation of λ DNA replication (4,5) and that eucaryotic Hsp70 proteins play in uncoating of clathrin vesicles (14) and in protein secretion (3,2). It will be of interest whether any of these postulated interactions of the DnaK and GroE proteins with cellular proteins are specific to their host or whether they are generic functions of these highly evolutionarily conserved proteins.

ACKNOWLEDGMENTS

We are grateful to C. Georgopoulos and J. Lutkenhaus for generously providing plasmids and phages. This research was supported by Public Health Service grant GM28988 from the National Institute of General Medical Sciences. B. B. was supported by a fellowship of the Deutscher Akademischer Austauschdienst, C. E. D. was supported by grant PF-3017 of the American Cancer Society.

LITERATURE CITED

1. Baker T.A. Grossman A.D. Gross C.A. (1984). A gene regulating the heat shock response in *E. coli* also affects proteolysis. Proc. Natl. Acad. Sci. USA 81:6779.
2. Chirico W.J. Waters M.G. Blobel G. (1988). 70 K heat shock related proteins stimulate protein translocation into microsomes. Nature 332:805.
3. Deshaies R.J. Koch B.D. Werner-Washburne M. Craig E.A. Schekman R. (1988). A subfamily of stress proteins facilitates translocation of secretory and mitochondrial precursor polypeptides. Nature 332:800.
4. Echols H. (1986). Multiple DNA-protein interactions governing high-precision DNA transactions. Science (USA) 233:1050.
5. Friedman D.I. Olson E.R. Georgopoulos C. Tilly K. Herscowitz I. Banuett F. (1984). Interactions of bacteriophage and host macromolecules in the growth of bacteriophage λ. Microbiol. Rev. 48:299.
6. Grossman A.D. Taylor W.E. Burton Z.F. Burgess R.R. Gross C.A. (1985). Stringent response in *Escherichia coli* induces expression of heat shock proteins. J. Mol. Biol. 186:357.
7. Itikawa H. Ryu J.-I. (1979). Isolation and characterization of a temperature-sensitive *dnaK* mutant of *Escherichia coli* B. J. Bacteriol. 138:339.
8. Krueger J.H. Walker G.C. (1984). *groEL* and *dnaK* genes of *Escherichia coli* are induced by UV irradiation and nalidixic acid in an *htpR*$^+$-dependent fashion. Proc. Natl. Acad. Sci. USA 81:1499.
9. Marsh L. Walker G.C. (1985). Cold sensitivity induced by overproduction of UmuDC in *Escherichia coli.* J. Bacteriol. 162:155.
10. McMullin T.W. Hallberg R.L. (1988). A highly evolutionarily conserved mitochondrial protein is structurally related to the protein encoded by the *Escherichia coli groEL* gene. Mol. Cell. Biol. 8:371.
11. Neidhardt F.C. VanBogelen R.A. Vaughn V. (1984). The genetics and regulation of heat shock proteins. Ann. Rev. Genet. 18:295.
12. Neidhardt F.C. VanBogelen R.A. (1987). in Neidhardt R.C. Ingraham J.L. Low K.B. Magasanik B. Schaechter M. Umbarger H.E. ed. *Escherichia coli* and *Salmonella typhimurium* cellular and molecular biology. p.1334.
13. Paek K.-H. Walker G.C. (1987). *Escherichia coli dnaK* null mutants are inviable at high temperature. J. Bacteriol. 169:283.
14. Rothman J.E. Schmid S.L. (1986). Enzymatic recycling of clathrin from coated vesicles. Cell 46:5.
15. Sakakibara Y. (1988). The *dnaK* gene of *Escherichia coli* functions in initiation of chromosome replication. J. Bacteriol. 170:972.
16. Tilly K. McKittrick N. Zylicz M. Georgopoulos C. (1983). The *dnaK* protein modulates the heat-shock response of *Escherichia coli*. Cell 34:641.

17. Tsuchido T. VanBogelen R.A. Neidhardt F.C. (1986) Heat-shock response in *Escherichia coli* concerns cell division. Proc. Natl. Acad. Sci USA 83:6959.
18. Wada M. Itikawa H. (1984). Participation of *Escherichia coli* K-12 *groE* gene products in the synthesis of cellular DNA and RNA. J. Bacteriol. 157:694.
19. Walker G.C. (1984). Mutagenesis and inducible responses to DNA damage in *Escherichia coli.* Microbiol. Rev. 48:60.
20. Ward J.E. Lutkenhaus J. (1985). Overproduction of FtsZ induces minicell formation in *E. coli.* Cell 42:941.

Stress-Induced Proteins, pages 37–47

THE ROLE OF THE ESCHERICHIA COLI HEAT SHOCK PROTEINS IN BACTERIOPHAGE LAMBDA GROWTH[1]

C. Georgopoulos[2], K. Tilly[3], D. Ang[2], G. N. Chandrasekhar[2,5], O. Fayet[2,6], J. Spence[2,7], T. Ziegelhoffer[2], K. Liberek[4], and M. Zylicz[4]

[2]Department of Cellular, Viral and Molecular Biology, University of Utah Salt Lake City, Utah 84132

[3]Laboratory of Biochemistry, NCI-NIH, Bethesda, Maryland

[4]Division of Biophysics, Department of Molecular Biology University of Gdansk, 24 Kladki, 80-822 Gdansk, Poland

ABSTRACT Bacteriophage λ relies heavily on the heat shock proteins of *E. coli* for its growth. In a purified system we have shown that the dnaK, dnaJ and grpE heat shock proteins are required for the initiation of λ DNA replication from *ori*λ. Their specific role appears to be the removal of λP from the protein complex assembled at *ori*λ. The groES and groEL heat shock proteins are required for the correct assembly of the λ prohead, specifically at the level of the formation of the λB dodecamer that constitutes the head-tail connector. Following λ infection the heat shock response is transiently turned on. We show that the σ^{32} polypeptide responsible for heat shock-specific transcription is an extremely unstable protein. The σ^{32} half life is stabilized both following λ infection and in *dnaK*756 mutant bacteria, thus allowing higher rates of heat shock gene transcription.

INTRODUCTION

Shortly after infection of *E. coli* by bacteriophage λ, the double-stranded circular λ DNA starts to replicate from its *ori*λ sequence (reviewed in references 1 and 2). Initiation of λ DNA replication is absolutely dependent on the presence of the λO and λP initiator proteins, an *ori*λ sequence and many host factors, i.e., dnaB, dnaG, dnaJ, dnaK, grpE, ssb, RNA polymerase, HU protein, DNA gyrase, ligase and DNA polymerase III. The dnaK, dnaJ and grpE proteins have been shown to belong to the

[1]This work was supported by grants from NIH and the Polish Academy of Sciences.

[5]Present address: Dept. of Microbiology, Columbia Univ., NY, NY 10032.
[6]Present address: CNRS, 118 route de Narbonne, 31062 Toulouse, France.
[7]Present address: Dept. of Biology, M.I.T., Cambridge, MA 02139.

so-called heat shock class of proteins (2,3,4). The rate of synthesis of the heat shock proteins is controlled at the transcriptional level and promoted by the σ^{32} polypeptide, the product of the *rpoH* gene (5).

ROLE IN λ DNA REPLICATION

The crude fraction II system, capable of *in vitro* replication of *oriC* plasmid DNA (6), was adapted to λ*dv* DNA replication and used to establish a biological assay for the purification of both phage- and host-coded factors required for λ DNA replication. For example, fraction II extracts prepared from *dnaJ*259 mutant bacteria will not replicate λ*dv* plasmid DNA *in vitro* unless provided with wild type dnaJ protein. The purification of the dnaJ protein is thus monitored through its ability to complement such a mutant fraction II extract in replication of λ*dv* plasmid DNA. Some of the protein factors shown in Fig. 1 were purified in this manner. The *dnaK* gene product is a 70,000-Mr polypeptide, an acidic protein with strong sequence homology to the eukaryotic hsp70 proteins (7). It possesses both a 5' nucleotidase activity and an autophosphorylation activity (8). It can be purified readily on an ATP-agarose column, following the same procedure described for the purification of hsp70-related proteins from HeLa cells (9). The *dnaJ* gene product is a highly basic, 37,000-Mr polypeptide, found almost exclusively in the membrane fraction and capable of binding to both

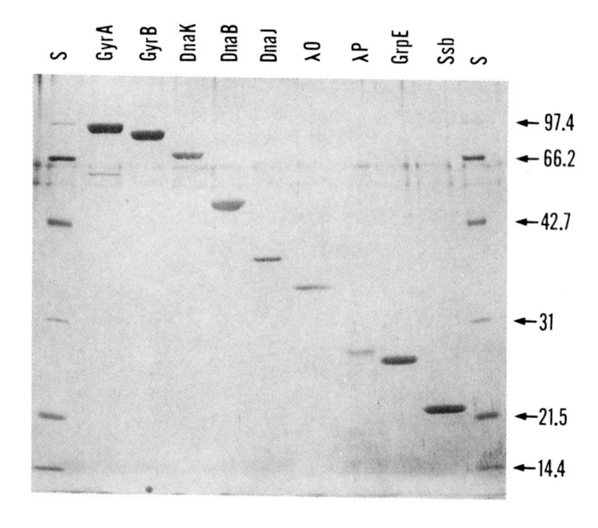

FIGURE 1. Purified *Escherichia coli* and bacteriophage λ replication proteins involved in the initiation of replication of λ DNA. Silver-stained 12 % SDS-polyacrylamide gel.

single-stranded and double-stranded DNA (10). The *grpE* gene product is an acidic 24,000-Mr polypeptide (11) which interacts with the dnaK protein in a salt resistant (i.e., hydrophobic) manner. The dnaK-grpE complex can be disrupted in the presence of ATP (12).

To define the role of the heat shock proteins in the replication of λ*dv* plasmid DNA, we reconstituted an *in vitro* purified protein system (M. Zylicz, K. Liberek, D. Ang, C. Georgopoulos, manuscript in preparation). Fig. 1 shows the highly purified prepriming proteins used in this system, which include the phage-coded λO and λP proteins, dnaB, dnaG, dnaJ, dnaK, grpE, ssb, DNA gyrase and DNA polymerase III holoenzyme. A similar purified λ*dv* replication system has been developed in the laboratory of Dr. Roger McMacken (13). The prepriming events, occurring before RNA primer formation by the dnaG primase can be subdivided into the four separable steps listed in Table 1. The first step is the binding of λO protein specifically to the *ori*λ DNA site. The second step is the addition of the already formed λP-dnaB complex to λO-*ori*λ. This step is carried out through the demonstrated λO and λP interaction (14). The third step is the disassembly and removal of the λP protein by the combined action of the dnaK and dnaJ proteins and ATP hydrolysis. DNA replication intermediates were isolated in the excluded volume following chromatography on a Sepharose 4B column according to the procedure described in reference 15. In order to determine the protein composition of the active replication complex, the λO, λP, and dnaK proteins were labeled *in vivo* with a ^{14}C-amino acid mixture and purified to homogeneity. The presence of the ^{14}C-labeled replication proteins in the intermediates listed in Table 1 was estimated and the results shown in Table 2.

TABLE 1

PREPRIMING STEPS IN λ DNA REPLICATION IN VITRO

	% recovery of DNA replication*
1. Binding of λO protein to *ori*λ sequence	76
2. Formation of an *ori*λ-λO-λP-dnaB complex	67
3. "Rearrangement" of *ori*λ-λO-λP-dnaB complex following action of the dnaK and dnaJ proteins	30
4. Unwinding of the double-stranded DNA by the dnaB helicase after addition of ssb, gyrase proteins and ATP to above intermediate.**	?

* following chromatography on a Sepharose 4B column.

** this step occurs simultaneously with dnaG-dependent RNA primer synthesis, provided dnaG protein and ribonucleotide triphosphates are added.

TABLE 2

Replication protein added to *ori*λ DNA	% of ^{14}C-λP or ^{14}C-λO in complex with λ*dv* plasmid DNA
1. ^{14}C-λO	96
2. ^{14}C-λO, λP, dnaB	94
3. ^{14}C-λO, λP, dnaB, dnaJ, dnaK	72
4. λO, ^{14}C-λP	10
5. λO, ^{14}C-λP, dnaB	63
6. λO, ^{14}C-λP, dnaB, dnaJ	70
7. λO, ^{14}C-λP, dnaB, dnaK	54
8. λO, ^{14}C-λP, dnaB, dnaJ, dnaK	8

The results presented in Table 2 suggest that both the dnaJ and dnaK heat shock proteins are responsible for the release of most of the λP, and some of the λO protein, from the preprimosomal complex (16). We were able to show that the λP release requires Mg^{+2} and the hydrolysis of ATP catalyzed by the dnaK protein (16). We propose that release of the highly hydrophobic λP protein from the *ori*λ-λO-λP-dnaB complex triggers the initiation of λ DNA replication by freeing the dnaB protein, thus allowing it to unwind nearby DNA through its helicase activity (17).

When ^{14}C-dnaK protein was used in similar experiments, little or no dnaK protein was present in the complex with DNA following dnaK and dnaJ protein action. The presence of the dnaJ protein in the replication intermediate was monitored by Western blot analysis. Most of the dnaJ protein was still found with the DNA complex, but it is not clear whether it is found there through its involvement with other replication proteins or is simply bound to DNA nonspecifically. Preliminary experiments also suggest that a small fraction of the dnaB is released from the preprimosomal complex, following dnaK and dnaJ protein action.

In the experiments presented in Table 2, the influence of grpE protein was not tested. The number of dnaK protein molecules needed to release λP protein was estimated to be approximately 900 per one copy of *ori*λ DNA sequence. This number should be compared to approximately 40-60 monomer equivalents for the rest of the proteins in the preprimosomal complex. An unwinding *ori*λ DNA-dependent assay based on the helicase activity of the dnaB protein was developed in order to quickly monitor the conversion of the inactive *ori*λ-λO-λP-dnaB intermediate to a competent replicating form (M. Zylicz, D. Ang, K. Liberek, C. Georgopoulos; manuscript in preparation). Following preincubation with the prepriming λ replication proteins, the extent of supercoiled λ*dv* DNA unwinding was monitored by electrophoresis in 1% agarose gels containing chloroquine, according to the procedure developed for the *oriC* system (18). When the replication complex shown on line 8, Table 2, was supplemented with ssb, DNA gyrase, and an ATP-regeneration system, extensive unwinding was observed (results not shown). When the concentration of dnaK protein was reduced fifteen times, no unwinding of λ*dv* plasmid DNA was observed. However, extensive unwinding was restored when grpE protein was added (Fig. 2). The grpE protein did not substitute entirely for either the dnaK or dnaJ requirement in this reaction: it simply reduced the need for large amounts of dnaK protein. In the presence of grpE protein, the requirement for complete DNA unwinding is approximately 50 monomers each of dnaK, dnaJ and grpE. The requirements for all other components in the complete system are the same both in the absence (high dnaK levels) and presence (low dnaK levels) of grpE protein. The results presented in Fig. 2 show that (a) grpE protein is required in a prepriming step either before or during unwinding of λ*dv* DNA, and (b) grpE protein somehow lowers the amount of dnaK protein required in the prepriming reaction. The precise mechanism of this grpE action is not known. What is known is that the dnaK and grpE proteins ultimately interact and that this interaction is extremely hydrophobic (i.e., resistant to high concentrations of salt) and is disrupted

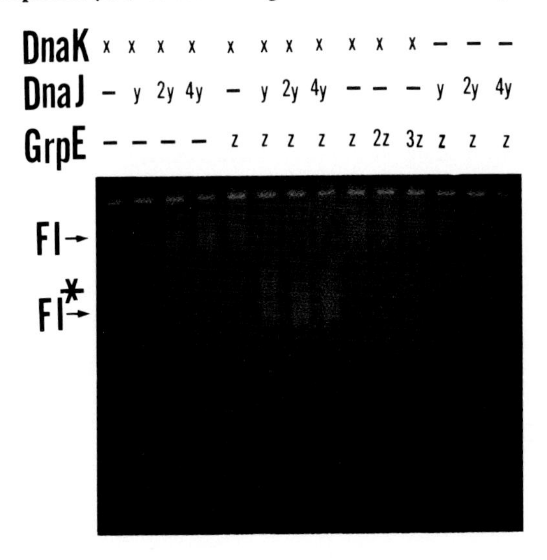

FIGURE 2. Agarose gel electrophoresis of supercoiled λ*dv* plasmid DNA (FI) and an extensively unwound form (FI*) generated during the prepriming stage of the initiation of λ DNA replication. The reactions (25 µl) contained: 40 mM Hepes/KOH pH 7.5, 7.2 mM magnesium acetate, 50 mM KCl, 30 mM NaCl, 2 mM ATP, 300 ng λ*dv* plasmid DNA, 71 ng λO, 150 ng λP, 150 ng dnaB, 860 ng ssb, 200 ng gyrA, 90 ng gyrB, 20 mM phosphocreatine, 500 ng creatine kinase, 0.4 mg/ml BSA, and various concentrations of: dnaK (x = 200 ng), dnaJ (y = 100 ng), grpE (z = 50 ng). The reaction was incubated 30 min at 38°C and stopped by addition of EDTA and SDS as described in (18).

in the presence of ATP (12). Various possibilities for the mode of action of the grpE protein in the prepriming steps include (a) direct participation in the λP protein release reaction in a manner analogous to dnaK and dnaJ, (b) alteration of the conformation of dnaK such that it becomes more efficient in releasing λP protein, (c) sequestration of the released λP protein in solution so that it cannot interfere with the dnaB protein's helicase activity, and (d) sequestration of λP protein at *ori*λ. This sequestration of λP at *ori*λ would allow dnaB to function, yet potentially accelerate the rate of formation of the next preprimosomal complex at *ori*λ.

ROLE IN λ MORPHOGENESIS

The *groE* heat shock genes of *E. coli* constitute an operon which maps at 94 min on the map, the order being promoter-*groES*-*groEL* (2). The two genes have been sequenced (19) and their products purified (20,21,22). The groEL protein is a decatetramer (composed of 57,000-Mr subunits) with seven-fold axis of symmetry and the groES protein is a 6-8 oligomer (composed of 10,000-Mr subunits).

The two proteins interact both *in vivo* and *in vitro* (22). Recently, the groEL protein of *E. coli* has been shown to be 46% identical at the amino acid sequence level to the plant Rubisco-binding protein (19). Furthermore, it also appears to be structurally and immunologically related to a mitochondrial-localized heat shock protein in *Tetrahymena* (23). The requirement for the groE morphogenetic protein in λ growth has been pinpointed to be at the level of assembly of the λ head-tail connector structure (24 and Fig. 3). The λ head-tail connector allows the proper assembly of the λ prohead, so that it can encapsidate λ DNA and join to λ tails to give rise to a mature phage (2). Although the exact role of the groE proteins in *E. coli* physiology is not known, it has been shown that excess groE proteins can suppress the temperature-sensitive phenotype of certain *dnaA*$^-$ mutations (25). It could be that the groE proteins perform assembly and disassembly functions analogous to that which they perform for λ head assembly. Whatever their function, we have recently shown that both the *groES* and *groEL* genes are essential for *E. coli* growth (26).

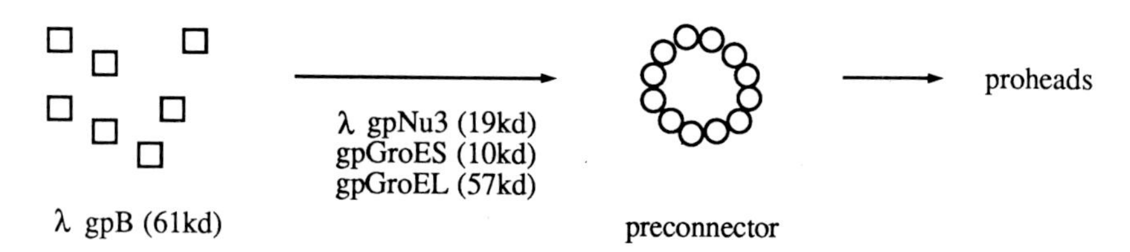

FIGURE 3. The assembly of the λ head-tail connector, consisting of a ring of 12 λ gpB subunits. The hypothetical pathway is derived both from genetic and biochemical studies.

PHAGE λ CONTROL OF HEAT SHOCK RESPONSE

The heat shock response in *E. coli* has been shown to be controlled at the transcriptional level by the σ^{32} polypeptide (5). *E. coli* σ^{32} null mutants cannot grow at temperatures above 20° (27). It has been shown by other laboratories and us that the σ^{32} polypeptide intracellular levels modulate the heat shock response (28). We have cloned the *rpoH* gene (which codes for σ^{32}) under the control of the *tac* inducible promoter. This plasmid, pCG179, allows massive synthesis of σ^{32} polypeptide upon IPTG induction (Fig. 4). By pulse-labeling for one min with ^{35}S-methionine, and chasing with cold methionine for various lengths of time we have shown that the intracellular half-life of σ^{32} can be as short as 30 sec at 42° (Fig. 4; 30-60 sec

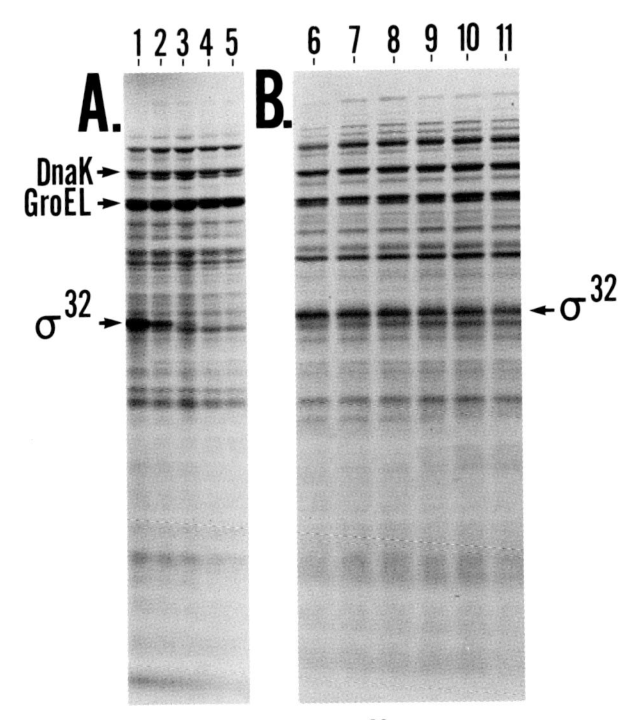

FIGURE 4. The intracellular half-life of σ^{32} is very short. RB791 (pCG179) bacteria growing in M9 medium were induced with IPTG (10^{-3} M) for 15 min. Subsequently they were pulse labeled for 1 min with 10 μCi ^{35}S-methionine (lanes 1 and 10. The cultures were subsequently chased with 100 μg/ml of cold methionine for 1 min (lanes 2 and 7), 2 min (lanes 3 and 8), 5 min (lanes 4 and 9), 10 min (lanes 5 and 10) and 20 min (lane 11). The proteins were separated following SDS-PAGE. Experiment in panel A (lanes 1-5) was done at 42°; panel B (lanes 6-11) at 22°.

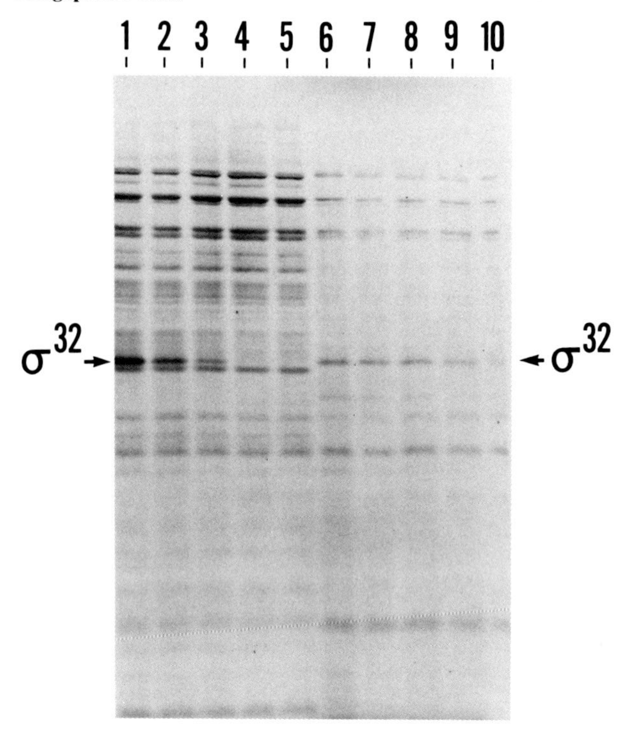

FIGURE 5. The half-life of σ^{32} is stabilized in *dnaK*756 bacteria. Isogenic RB791 (lanes 1-5) and RB791 *dnaK*756 (lanes 6-10) bacteria, carrying the CG179 plasmid, were grown in M9 medium at 37°, induced with IPTG, labeled and chased as described in the legend to Fig. 4. One min pulse (lanes 1 and 6), 1 min chase (lanes 2 and 7), 2 min chase (lanes 3 and 8), 5 min chase (lanes 4 and 9) and 10 min chase (lanes 5 and 10).

variability from experiment to experiment), approximately 60-90 sec at 37° and 30° (results not shown), and as long as 15 min at 22°C (Fig. 4).

Previous results from our laboratory showed that *dnaK*756 mutant bacteria overproduce heat shock proteins at 30° and 37° and fail to properly turn off the heat shock response at 42° (29). Fig. 5B shows that the half-life of the σ^{32} polypeptide at 37° is increased at least five-fold in *dnaK*756 bacteria relative to their isogenic parent (Fig. 5A). This increased stability of σ^{32} polypeptide is most likely responsible for the abnormal heat shock response in *dnaK*756 bacteria.

Following bacteriophage λ*c*I857*cro*27 prophage induction at 42°, the synthesis of early λ proteins continues unabatedly, host protein synthesis is shut off, but heat shock protein synthesis continues (30 and Fig. 6A). Under these conditions the half-life of

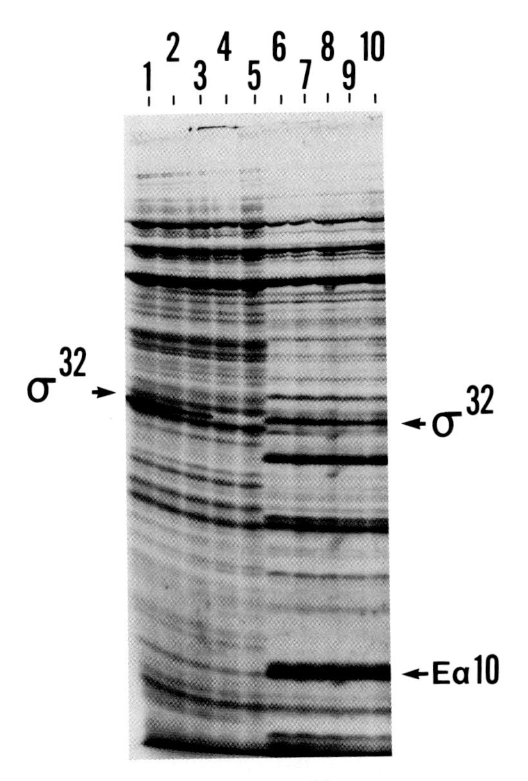

FIGURE 6. Phage λ infection stabilizes σ^{32} half-life. RB791 (lanes 1-5) and RB791 (λ*c*I857*cro*27*Sam*7) (lanes 6-10) bacteria, carrying the pCG179 plasmid, were grown at 30° in M9 medium. Subsequently they were shifted to 42° and IPTG was simultaneously added at 10^{-3}M. At 15 min after the shift, the cultures were labeled and chased as described in the legend to Fig. 4. One min pulse (lanes 1 and 6), 1 min chase (lane 2 and 7), 2 min chase (lanes 3 and 8), 5 min chase (lanes 4 and 9), and 10 min chase (lanes 5 and 10).

the σ^{32} polypeptide is also stabilized at least five-fold (Fig. 6B). The ability of bacteriophage λ to modulate the heat shock response under these conditions has been traced to the *c*III gene. It has been shown that the overproduction of λcIII protein from an inducible promoter is both necessary and sufficient to (a) turn on the heat shock response in *E. coli*, even at 30°, and (b) stabilize the half-life of the σ^{32} polypeptide (31).

CONCLUSIONS

Bacteriophage λ relies on the heat shock response of *E. coli* for both its DNA replication and the morphogenesis of its head. The dnaK, dnaJ and grpE proteins have been shown to act synergistically to release the λP protein from the λO-λP-dnaB protein complex assembled at *ori*λ. The λP protein, through its interaction with both λO and dnaB, allows the localization of dnaB at *ori*λ. However, once the protein complex has assembled, λP becomes inhibitory. Thus, the dnaK, dnaJ and grpE proteins are needed to release λP with the help of a hydrolyzable triphosphate. Clearly these three proteins assist in the disassembly of a complex. Without these heat shock proteins, λ DNA does not replicate, even for one cycle.

The mode of action of the *E. coli* groES and groEL proteins appears to be the assembly of the λ head-tail connector made up of a λB dodecamer. The groE proteins themselves are not part of the finished structure. This "chaperone" role of the groE proteins is analogous to that seen with the groEL-analogue, the plant Rubisco-binding protein(19).

Bacteriophage λ not only relies heavily on the heat shock proteins of *E. coli* for its growth, but it also positively modulates the heat shock response following infection. This is accomplished chiefly through the action of the λcIII regulatory protein. The apparent mechanism seems to be the stabilization of the short half-life of σ^{32} polypeptide, leading to higher levels of heat shock protein production.

REFERENCES

1. Furth ME, Wickner SH (1983). In Hendrix RW et al. (eds). "Lambda II," New York: Cold Spring Harbor Laboratory, p.145.
2. Friedman DI, Olson ER, Tilly K, Georgopoulos C, Herskowitz I, Banuett F (1984). Microbiol. Rev. 48:299.
3. Georgopoulos C, Tilly K, Drahos D, Hendrix R (1982). J. Bacteriol. 149:1175.
4. Bardwell JCA, Tilly K, Craig E, King J, Zylicz M, Georgopoulos C (1986). J. Biol. Chem. 261:1782.
5. Grossman AD, Erickson JW, Gross CA (1984). Cell 38:383.
6. Fuller RS, Kaguni JM, Kornberg A (1981). Proc. Natl. Acad. Sci. USA 78:7370.
7. Bardwell JCA, Craig EA (1984). Proc. Natl. Acad. Sci. USA 81:848.
8. Zylicz M, LeBowitz JH, McMacken R, Georgopoulos C (1983). Proc. Natl. Acad. Sci. USA 80:6431.
9. Welch WJ, Feramisco JR (1985). Mol. Cell. Biol. 5:1229
10. Zylicz M, Yamamoto T, McKittrick N, Sell S, Georgopoulos C (1985) J. Biol. Chem. 260:7591.
11. Ang D, Chandrasekhar GN, Zylicz M, Georgopoulos C (1986) J. Bacteriol. 167:25.
12. Zylicz M, Ang D, Georgopoulos C (1987). J. Biol. Chem. 262:17437.

13. McMacken R, Alfano C, Gomes B, LeBowitz JH, Mensa-Wilmot K, Roberts JD, Wold M (1987). In McMacken R, Kelly T (eds): "Mechanisms of DNA replication and recombination," New York: Alan R. Liss, p.227.
14. Zylicz M, Gorska I, Taylor K, Georgopoulos C (1984). Mol. Gen. Genet. 196:401.
15. Wickner SH, Zahn K (1986). J. Biol. Chem. 261:7537.
16. Liberek K, Georgopoulos C, Zylicz M (1988). Proc. Natl. Acad. Sci. USA, in press.
17. LeBowitz JH, McMacken R (1986). J. Biol. Chem. 261:4738.
18. Baker TA, Sekimizu K, Funnell BE, Kornberg A (1986) Cell 45:53.
19. Hemmingsen SM, Wollford C, van der Vies SM, Tilly K, Dennis DT, Georgopoulos CP, Hendrix RW, Ellis RJ (1988). Nature 333:330.
20. Hendrix RW (1979). J. Mol. Biol. 129:375.
21. Hohn T, Hohn B, Engel A, Wurtz M, Smith PR (1979). J. Mol. Biol. 129:359.
22. Chandrasekhar GN, Tilly K, Woolford C, Hendrix R, Georgopoulos C (1986). J. Biol. Chem. 261:12414.
23. McMullin TW, Hallberg RL (1988). Mol. Cell. Biol. 8:371.
24. Kochan J, Murialdo H (1983). Virology 131:100.
25. Fayet O, Louarn J-M, Georgopoulos C (1986). Mol. Gen. Genet. 202:435.
26. Fayet O, Ziegelhoffer T, Georgopoulos C (1988). Submitted to J. Bacteriol.
27. Zhou Y-N, Kusukawa N, Erickson JW, Gross CA, Yura T (1988). J. Bacteriol., in press.
28. Strauss DB, Walter WA, Gross CA (1987). Nature 329:348.
29. Tilly K, McKittrick N, Zylicz M, Georgopoulos C (1983). Cell 34:641.
30. Tilly K, Chandrasekhar GN, Zylicz M, Georgopoulos C. (1985). In Leive L (ed): "Microbiology 1985," Washington DC: American Society for Microbiology, p322.
31. Bahl H, Echols H, Straus D, Court D, Crowl R, Georgopoulos CP (1987). Genes and Devel. 1:57.

B: EUCARYONS

Stress-Induced Proteins, pages 51–61

COMPLEX REGULATION OF THREE HEAT INDUCIBLE HSP70 RELATED GENES IN *Saccharomyces cerevisiae*[1]

Elizabeth Craig, William Boorstein, Hay-Oak Park, David Stone, and Charles Nicolet

Department of Physiological Chemistry, University of Wisconsin-Madison, Madison, Wisconsin 53706

ABSTRACT Expression of the *SSA1*, *SSA3*, and *SSA4* genes of *S. cerevisiae* is induced upon temperature upshift. Heat shock elements (HSE's) which are able to activate a heterologous promoter are present in the upstream region of each of the genes. The basal level of expression regulated by an HSE in the *SSA1* promoter is modulated through the action of an upstream repressing sequence (URS). In addition, expression of the *SSA1* gene is responsive to the level of *SSA1* protein, implying self-regulation of the *SSA1* gene.

INTRODUCTION

The 70 kDa heat shock genes comprise the largest known multigene family transcribed by RNA polymerase II in *S. cerevisiae*. The maintenance of more than eight related genes is likely to result from the requirement for differential regulation of proteins with nearly identical functions as well as for proteins which carry out related but distinct functions. Mutational analysis in which various combinations of members of this related family were inactivated demonstrates at least partial functional equivalence amongst groups of these genes (1,2). Furthermore, it has been shown that while strains bearing certain hsp70 mutations are inviable, the viability can be restored by altering the transcriptional regulation of the remaining genes (3). Thus, the differential ability of the HSP70

[1]This work was supported by a grant from the National Institutes of Health (GM 31107) to E.A.C. and the Lucille P. Markey Charitable Trust, Miami, Florida (H.O.P.).

genes to compensate for each other has been shown to be, at least in part, due to differences in regulation.

Our studies of three of the heat inducible genes related to HSP70 (SSA1, SSA3, and SSA4) indicate that while common mechanisms are likely to give rise to the heat induction, there are important differences between promoter regions and regulatory circuits which result in distinct expression patterns. Some cis- and trans-acting regulatory components of three of the heat inducible yeast HSP70 genes are addressed here.

RESULTS AND DISCUSSION

Heat Shock Elements

The SSA1, SSA3, and SSA4 genes are induced by thermal stress. The 5' non-transcribed region of each of these genes contains multiple occurrences of sequences which are closely related to the heat shock element consensus sequence (4). HSE's have been shown to be binding sites of a positively acting heat shock transcription factor (HSF) and facilitate transcription in an *in vitro* system (5). The diamonds in Figure 1 represent the positions of HSE's in the promoters of the SSA1, SSA3, and SSA4 genes. Each of the genes has at least three such regions within 350 nucleotides of the transcriptional initiation sites. These genes also contain an overlapping arrangement of HSE's similar to that found in the 5' flanking region of *Drosophila* HSP83 and several other heat inducible genes.

Extensive 5' flanking regions of SSA1, SSA3, and SSA4 were fused to the *lacZ* gene of *E. coli* to facilitate the study of expression driven from these promoters. The expression of these fusions under different culture conditions, as assayed by ß-galactosidase activity, parallels that of the native genes. Deletion analysis has demonstrated that HSE's are required for the induction of these genes in response to thermal stress. However, all of the HSE-like sequences are not essential for heat inducible expression. Sequences that were shown to be functionally important by 5' deletion experiments, underlined in Figure 1, were studied further.

Synthetic oligonucleotides identical to the HSE sequences of these 3 genes have been tested for their ability to act as heat inducible upstream activating sequences, UAS^{HS}'s. The fragments tested are shown in

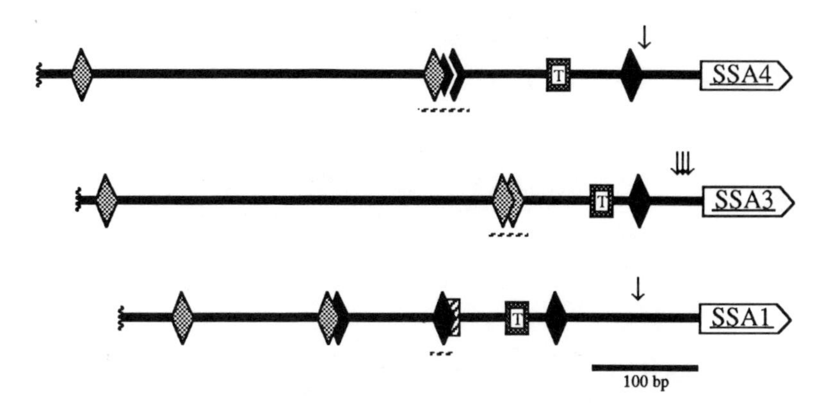

FIGURE 1. Schematic diagram of the 5' non-coding regions of SSA1, SSA3, and SSA4. The open boxes indicate the amino termini of the coding regions. Diamonds represent matches to the heat shock element consensus sequence, CNNGAANNTTCNNG. Black diamonds indicate exact matches and single base mismatches to this consensus. Grey diamonds mark sequences which differ from the consensus by two nucleotides. The boxes containing the letter T represent occurrences of the cannonical TATA sequence. Vertical arrows represent the major transcriptional initiation sites. The striped box marks the URS sequence present upstream of the SSA1 gene. The sequences of the underlined regions are shown in Figure 2.

Figure 2. These sequences were cloned into a cytochrome c, CYC1, lacZ fusion vector in place of the UAS^{CYC1}. All three sequences were able to activate the CYC1 downstream promoter elements in a heat inducible manner and thus constitute functional UAS^{HS}'s.

The SSA1, SSA3, and SSA4 oligonucleotides which are sufficient to confer heat inducible transcriptional regulation each contain regions very similar to the HSE consensus sequence. Matches to the palindromic heat shock elements, CNNGAANNTTCNNG, are underlined in Figure 2. As indicated by the triangles, these sequences contain one or more additional GAA blocks which comprise the 'core' of the consensus sequence. These additional HSE-like sequences have been observed adjacent to the HSE's of a number of heat shock genes from widely divergent species. Recently these nucleotides were shown to be required for

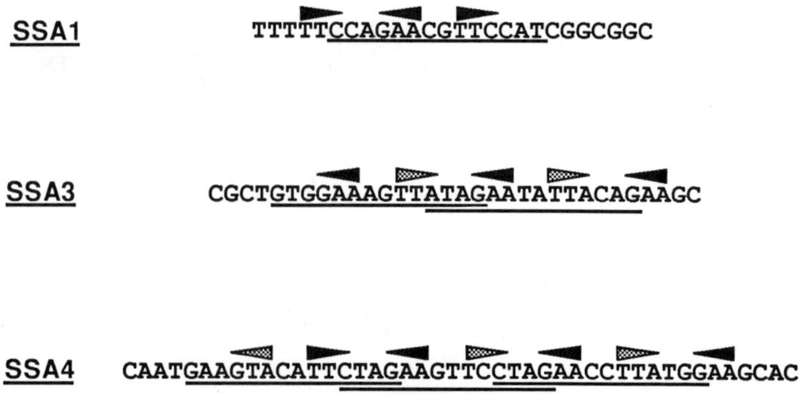

SSA1 TTTTTCCAGAACGTTCCATCGGCGGC

SSA3 CGCTGTGGAAAGTTATAGAATATTACAGAAGC

SSA4 CAATGAAGTACATTCTAGAAGTTCCTAGAACCTTATGGAAGCAC

FIGURE 2. Heat shock elements and flanking sequences from SSA1, SSA3, and SSA4. The bars indicate the positions of the consensus HSE's. The SSA1 HSE shown here is referred to as "HSE2". Triangles mark the component GAA repeats with 2 nucleotide relative spacing. Black triangles are over exact matches to GAA. Gray triangles are over sequences with a mismatch to the component repeat element.

HSE activity in a more rigorous functional test of the precise requirements for this heat inducible enhancer in Drosophila (6,7). The three functional HSE's shown here conform to the model proposed by Voellmy and colleagues, which postulates the requirement of at least 3 GAA blocks in alternating orientation within rigid spacing constraints (7). Limited experimental evidence supports similar expanded sequence requirements in S. cerevisiae. While the SSA1 oligonucleotide shown in Figure 2 is a functional UAS^{HS}, a smaller oligomer that includes the entire underlined consensus region, but not the 5' TTC sequence, has no UAS^{HS} activity (8). Reciprocal binding studies using a Drosophila heat shock transcription factor and yeast HSE's, and vice versa, further suggest the activation mechanisms and regulatory sequence requirements are very similar between these species (9).

Basal Level of Expression and Negative Regulation of UAS^{HS}

An HSE from the SSA1 promoter, denoted HSE2, was shown to be sufficient for heat inducible and basal level expression as measured from a lacZ reporter gene. Further tests of the sequences necessary for heat inducible expression led to the identification of a sequence involved in modulating the basal activity of HSE2 (8). The 26 bp fragment containing HSE2 and its immediately adjoining nucleotides (Figure 2), inserted in place of the CYC1 UAS, could confer a substantial basal level of expression to the heterologous reporter gene. However, when a larger DNA fragment (137 bp) containing the same HSE replaced UAS^{CYC1}, the basal, but not the heat inducible, expression was dramatically lower. The demonstrated inhibition of expression in this construction suggests that sequences present in the 137 bp DNA fragment are also negatively regulating the basal expression of SSA1.

To identify the sequences involved in the negative regulation, series of deletions were generated from both ends of the 137 bp DNA fragment. Deletions which lack DNA downstream of the UAS^{HS} showed an increased basal level of expression, indicating that these deletions destroy the upstream repressing sequence (URS). To test directly whether these HSE2 proximal sequences are able to cause inhibition of the basal level of expression, a 40 bp oligonucleotide containing HSE2 and the putative URS was synthesized and inserted in place of the CYC1 UAS. The similar low basal level of expression from the 40 bp and 137 bp fragments suggests that the URS is contained within this smaller region. The 40 bp fragment contains a 13 bp sequence that has two mismatches, in the inverted orientation, to a consensus sequence, $TAGCCGCCGPu_4$, proposed by T.G. Cooper and colleagues to be a negative control site in yeast (11). In the SSA1 promoter, the 13 bp sequence partially overlaps HSE2.

Since the 137 bp, 40 bp and 26 bp DNA fragments all confer similar levels of heat inducible expression, it is likely that the negative control site plays a role only in basal expression of the SSA1 gene. Although there are other HSE's in the SSA1 promoter (see Figure 1), the results of deletion analysis indicate that HSE2 plays a key role in the control of native SSA1 gene expression (8); the URS acts to inhibit HSE2 activity in the absence of an inducing stimulus. Similar results were obtained in a study of the human ß-interferon promoter which contains a constitutive transcription element and a negative regulatory sequence that prevents enhancer activity prior to induction (12,13). *In vivo* DNAse I footprinting suggests that repressor molecules bind to the ß-interferon

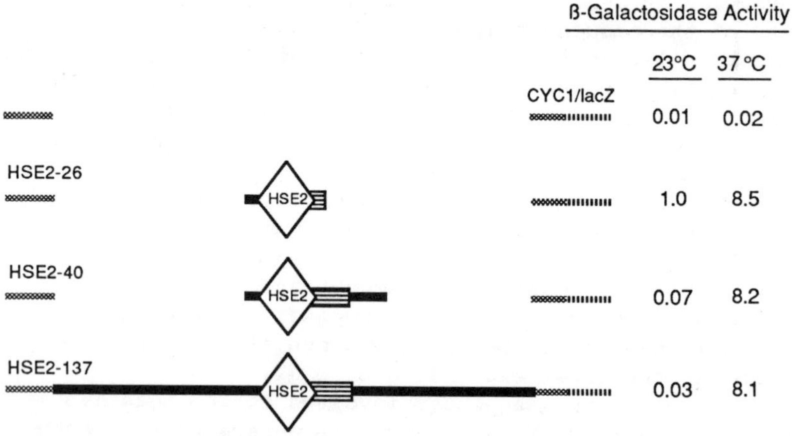

	ß-Galactosidase Activity 23°C	ß-Galactosidase Activity 37°C
CYC1/lacZ	0.01	0.02
HSE2-26	1.0	8.5
HSE2-40	0.07	8.2
HSE2-137	0.03	8.1

FIGURE 3. The activity of SSA1-CYC1 promoter fusions before and after heat shock. Segments of the SSA1 promoter were inserted into the CYC1 promoter-lacZ fusion vector (pLG670Z) (10) in place of the CYC1 UAS. The enzymatic activity of ß-galactosidase in yeast cells transformed with the constructions was determined during logarithmic growth at 23°C and 60 min after shift to 37°C. The ß-galactosidase activity of cells grown at 23°C containing the plasmid with the 26 bp segment was arbitrarily given the value of 1. All other ß-galactosidase levels are presented as a fraction of the activity present in cells containing this fusion.

regulatory region under non-inducing conditions. After induction the repressor molecules dissociate from the DNA and a positive transcription factor binds to and activates the ß-interferon promoter. It will be interesting to see if a factor binding to the URS prevents binding of the heat shock transcription factor prior to induction.

The SSA3 and SSA4 regulatory sequences do not contain elements similar to the URS, yet they are expressed at very low levels under non-inducing conditions. Preliminary evidence suggests that both of these genes have sequences which function to repress the basal levels of expression. Deletion of these regions results in three to four fold increases in basal expression. The hybrid promoter constructs containing the SSA1, SSA3, and SSA4 HSE oligonucleotides give rise to different levels of expression under non-inducing conditions, suggesting that the HSE sequences or their immediate context also determines the basal levels of expression.

Self Regulation of SSA1

Although there is some understanding of the induction of the heat shock response, little is known about how the response is shut off. The results of several studies in Drosophila melanogaster and Escherichia coli suggest that HSP70 is autoregulatory (14, 15). To test the autoregulation model in S. cerevisiae, we determined the effect of overproduction of Ssa1 protein on the activity of its own promoter. The constitutive level of Ssa1 was increased by fusing the SSA1 structural gene to the GAL1 promoter, thus putting Ssa1 production under the control of a conditional promoter. A reporter vector consisting of a SSA1-lacZ translational fusion was used to assess SSA1 promoter activity. In a strain containing a single copy of the GAL1-SSA1 fusion and producing approximately 3-fold the normal heat shock level of Ssa1, the basal and induced levels of ß-galactosidase were reduced 2-fold, but the relative increase in ß-galactosidase activity after heat shock was similar to that of the isogenic control strain. In a strain producing approximately 10-fold more than the normal heat shock level of Ssa1, induction of ß-galactosidase activity by heat shock was almost entirely blocked. High-level overexpression of Ssa1 did not affect either a CUP1-lacZ translational fusion or SSA2-lacZ translational fusion indicating that the decrease in expression of the SSA1 fusion was not a general effect on gene expression.

The observed effect on the expression of the SSA1-lacZ fusion could have been due to effects at the level of transcription or translation. That the effect is at least in part rendered at the transcriptional level was shown by testing the effect of overproduction of Ssa1 on some of the transcriptional SSA1-CYC1-lacZ fusions shown in Figure 3. Expression driven by the 137 bp fragment surrounding HSE2 was repressed when Ssa1 was overexpressed (Fig. 4). Since this fusion contains only a segment of the promoter region and no transcribed sequences, it demonstrates that the inhibition occurs at the level of transcription. Expression of the fusions containing the 26 bp fragment (HSE2) and the 40 bp fragment (HSE2 and URS) was not inhibited significantly. Therefore, we conclude that the 137 bp

DNA fragment contains a sequence which is responsive to the amount of Ssa1 in the cell, but the smaller fragments containing HSE2 and URS do not. This result implies that the self-regulation observed does not act through the heat shock transcription factor (HSF) since the inhibition is

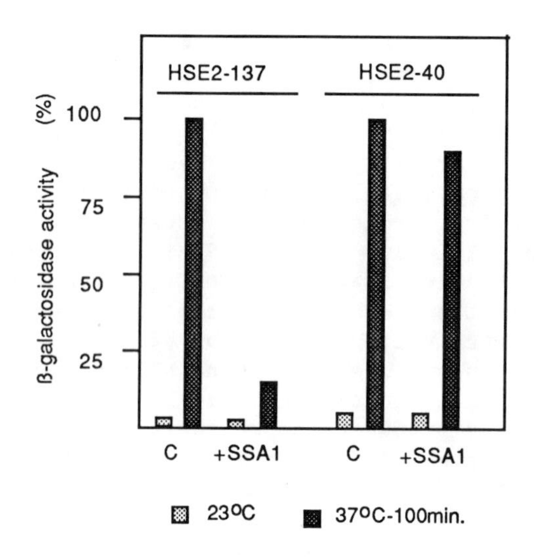

FIGURE 4. The activity of SSA1-CYC1 promoters in the presence and absence of excess Ssa1. The fusions HSE2-137 and HSE2-40 shown in Figure 3 were transformed into yeast strains containing a control plasmid (C) and a plasmid containing the GAL1 promoter linked to SSA1 (+SSA1). The ß-galactosidase activity of these transformants grown in galactose-based media at 23°C and 90 minutes after a shift to 37°C was determined. The level in control cells after heat shock was arbitrarily given the value of 100%.

only seen when sequences in addition to HSE's are present. Similarly, the repression by overproduction of Ssa1 does not seem to act through the URS. The self-regulating element could be independent of the other two elements or overlap them. Whether the observed effect of Ssa1 is a direct or an indirect one is not addressed by these experiments.

Possible trans-acting regulators. From the preceding discussion it should be clear that the expression of some, if not all, of the HSP70 related genes can be regulated by factors other than the binding of a single activator protein (HSF) to the appropriate target sequence (HSE). Other trans-acting factors interact with the promoter regions of these genes, and perhaps with the mRNAs themselves, to carry out a fine-tuning of the response of the heat shock genes under a variety of conditions. We designed a cloning strategy in an attempt to identify loci which affect the expression of heat shock genes. This strategy might be expected to identify the HSF gene, but it would not be limited to isolation of genes whose products interact with HSE's or even act at the level of transcription. The rationale for the approach is based on the expectation that overexpressed regulatory molecules may result in an exaggerated regulatory response. For example, by overexpressing an activator protein (by virtue of gene dosage), one might see an increase in the amount of expression of a reporter gene fused to a heat shock gene promoter, circumventing the normal regulation.

The experiment was set up as follows: yeast cells were first transformed with a centromeric vector containing the SSA4 promoter fused to the murine dihydrofolate reductase (DHFR) gene. DHFR confers resistance to the drug methotrexate (mtx), but cells containing this fusion are mtx sensitive since the SSA4 promoter is not active in unstressed cells. This strain was then transformed with a yeast genomic library constructed in a multicopy plasmid. The expression of genes cloned onto such a plasmid can be elevated 5-20 fold relative to the single genomic locus. Transformants were simultaneously selected for the nutritional marker on the plasmid and for mtx resistance. A number of colonies were isolated, and subjected to a screening procedure to eliminate transformants whose resistance wasn't due to increased expression from the SSA4 promoter.

One clone has been examined in the most detail. We have named the gene present on this plasmid SRL, for Stress Related Locus, since it appears to affect the expression of at least some of the heat shock genes in the HSP70 related family. SRL itself is a heat inducible gene. Furthermore, srl mutants are both heat sensitive and cold sensitive for growth. The SRL gene seems to exert its effects by controlling both mRNA and protein levels. The mechanisms by which these effects are achieved are still obscure. Isolation of this interesting gene points out the validity of the cloning approach utilized. The identification of SRL further demonstrates that a variety of loci can influence the expression of HSP70 related genes.

CONCLUSION

Members of the HSP70 related multigene family are expressed in a highly complex pattern that results from regulation at many levels. We have focused on a subset of the regulatory components of the heat inducible members of the SSA subfamily. The common stress response of SSA1, SSA3, and SSA4 results primarily from the stress responsive promoter element, the HSE, present in each of these genes. HSE sequences from SSA1, SSA3, and SSA4 have been shown to be sufficient to mediate heat inducible expression. Other aspects of expression from these genes are controlled in part via distinct cis-acting loci. The URS acts to maintain a low basal level of expression of the SSA1 gene. The SSA1 gene product is an example of a different type of regulator which acts in trans to inhibit either directly or indirectly its own expression. Future studies are required to elucidate the specificity of this autoregulation and whether the other SSA gene products act in a similar manner. SRL is another trans-acting regulator which affects expression of SSA4 in a positive manner. The diverse regulatory phenomena which we are studying have presumably arisen following duplication of a single HSP70 ancestral gene. Differential regulation can be seen as an evolutionary strategy which allows the cell to generate diversity in function by having related gene products present under differing environmental conditions.

REFERENCES

1. Craig EA, Jacobsen K (1984). Mutations of the heat-inducible 70 kilodalton genes of yeast confer temperature sensitive growth. Cell 38:841.
2. Craig EA, Jacobsen K (1985). Mutations in cognate genes of Saccharomyces cerevisiae hsp70 result in reduced growth rates at low temperatures. Mol. Cell. Biol. 5:3517.
3. Werner-Washburne M, Stone D, Craig EA (1987). Complex interactions among members of an essential subfamily of hsp70 genes in Saccharomyces cerevisiae. Mol. Cell. Biol. 7:2568.

4. Pelham HRB, Bienz M (1982). A synthetic heat shock promoter element confers heat-inducibility on the herpes simplex virus tk gene. EMBO 1:1473.
5. Parker CS, Topol J (1984). A Drosophila RNA polymerase II transcription factor contains a promoter-region-specific DNA-binding activity. Cell 36:357.
6. Xiao H, Lis JT (1988). Germline transformation used to define key features of heat-shock response elements. Science 239:1139.
7. Voellmy R. Personal Communication.
8. Slater M, Craig EA (1987). Transcriptional regulation of an hsp70 heat shock gene in the yeast Saccharomyces cerevisiae. Mol. Cell. Biol. 7:1906.
9. Wiederrecht G, Shuey D, Kibbe W, Parker C (1987). The Saccharomyces and Drosophila heat shock transcription factors are identical in size and DNA binding properties. Cell 48:507.
10. Guarente L, Ptashne M (1981). Fusion of Escherichia coli lacZ to the cytochrome c gene of Saccharomyces cerevisiae. Proc. Natl. Acad. Sci. USA 78:2199.
11. Sumrada RA, Cooper TG (1987). Ubiquitous upstream repression sequences control activation of the inducible arginase gene in yeast. Proc. Natl. Acad. Sci. USA 84:3997.
12. Goodbourn S, Burstein H, Maniatis T (1986). The human ß-interferon gene enhancer is under negative control. Cell 45:601.
13. Zinn K, Maniatis, T (1986). Detection of factors that interact with the human ß-interferon regulatory region in vivo by DNAase I footprinting. Cell 45:611.
14. DiDomenico BJ, Bugoisky GE, Lindquist SL (1982). The heat shock response is self-regulated at both the transcriptional and post-transcriptional levels. Cell 31:593.
15. Tilly K, McKittrick N, Zylicz M, Georgopoulos C (1983). The dnaK protein modulates the heat-shock response of Escherichia coli. Cell 34:641.

Stress-Induced Proteins, pages 63–72

TRANSCRIPTIONAL REGULATION OF THE HEAT SHOCK GENES BY CYCLIC-AMP AND HEAT SHOCK IN YEAST

Kazuma Tanaka, Takehiro Yatomi,
Kunihiro Matsumoto*, and Akio Toh-e

Department of Fermentation Technology,
Hiroshima University, Higashihiroshima 724,
Japan and *DNAX Research Institute, 901
California Avenue, Palo Alto, CA 94304-1104

ABSTRACT The involvement of the cAMP-pathway in transcriptional regulation of the heat shock genes was examined. From analysis of the expression of the UBI4 and SSA1 genes, we propose that cAMP regulates the expression of the heat shock genes, but does not mediate the heat shock response. Preliminary results of dissection of the heat shock promoters seems to support this hypothesis.

INTRODUCTION

The G0/G1 phase of eukaryotic cells has a crucial role in determination of initiation of the next round of the cell cycle. The study of gene regulation in this phase seems to be important for the fundamental understanding of the G0/G1 phase. In yeast, Saccharomyces cerevisiae, it was shown that the several heat shock proteins (HSPs) are synthesized in G0 arrested cells (1).

Recent work by Matsumoto et al. (2) clearly indicates that the reduction of the cAMP-dependent protein phosphorylation is a prerequisite for G0 arrest in S. cerevisiae. Thus, we supposed that the HSP genes may be regulated by cAMP-dependent protein kinase (A-kinase) in a negative fashion. Transcriptional analyses of the UBI4 and SSA1 genes, which encode polyubiquitin (3) and a hsp70 homolog (4), respectively, indicate that both genes are regulated by heat shock and by cAMP. Moreover, our results suggest that these divergent signals regulate HSP genes through different upstream elements.

METHODS

Strains. A yeast strain, AM18-5C-M1 (MAT α cyr1-1 ura3), a spontaneous ura3 mutant obtained from AM18-5C (5) by the 5-fluoro-orotic acid selection method (6), was used as a host for lacZ plasmids. The bcy1::URA3 mutant, RA1-13D-R (MATa ura3 leu2 trp1 his3 bcy1::URA3), was constructed from RA1-13D by disrupting the BCY1 gene with the URA3 gene (7).

Media. YPD medium contains 2 % glucose, 2 % polypeptone, and 1 % Yeast extract. SD medium is 2 % glucose and 0.7 % Yeast Nitrogen Base without amino acids (Difco) and, if required, is supplied with appropriate amino acids or bases at 20-40 µg/ml. SDA medium is SD medium containing 0.5 % Casamino Acids (Difco). SDAA medium is SDA medium supplemented with 0.5 mM cAMP.

Plasmids. Five lacZ plasmids were constructed by standard molecular biological techniques (8) for the analysis of heat sock promoters. pJU6 or pJU24 were made by inserting the UBI4-lacZ fusion genes into the centromeric DNA-based vector, YCp19-LEU2, in which the URA3 gene and the TRP1 gene of YCp19 (9) were replaced with the LEU2 gene, or into the 2 µm-based vector, YEp24 (10). The UBI4-lacZ genes of pJU6 and pJU24 are composed of UBI4 upstream sequences, the 237 bp UBI4 coding sequence, and 6.2 kb of lacZ DNA. The UBI4 upstream sequence of pJU6 is about 100 bp longer than that of pJU24, 767 bp. pJU24-HSE was obtained from pJU24 by deleting the 10 bp XbaI-XbaI DNA segment from the only exact copy of the 14 bp consensus heat shock box sequence (11) present in the UBI4 upstream region (12). The SSA1-lacZ fusion gene, constructed essentially as described previously (13), was cloned into YCp19 or YCp19-LEU2 and these plasmids were designated as pJU25 or pJU26. pZJ-HSE2-26 and pZJ-HSE2-26-R (14) were kindly provided by Dr. E. A. Craig.

lacZ assays for heat shock and cAMP responses. AM18-5C-M1 was transformed with various lacZ fusion plasmids. At least two transformants of the individual plasmids were cultured in SDAA medium into exponential growth phase. Cells were collected, washed twice with water and suspended in 0.7 ml of water. 0.2 ml samples were inoculated into two 5ml portions of SDAA and one 5ml portion of SDA media. One SDAA culture (control) and the SDA culture (-cAMP) were shaken for 12 hours at 25°C ; the other SDAA culture was shifted to 37°C for 2 hours after cultivation at 25°C for 10 hours (heat shock). Then, β -galactosidase activities were measured essentially as described (15) using cells permeabilized with chloroform and sodium dodecyl sulfate as enzyme sources. A660 values did not exceed 0.5 at time of harvest.

Measurement of thermal tolerance. Cells grown exponentially at 25°C were pre-incubated at 25° C (control) or 37°C (heat shock) for 1 hour. Samples were taken, diluted appropriately with water and plated on YPD. Each remaining culture was thermal treated at 52°C for 8 min and cooled on ice. Thermal treated samples were also diluted and plated on YPD. After 3 days incubation at 30°C, the % survival was determined by dividing the viable count of the thermal treated culture by that of the control.

Other techniques. Northern hybridizations were as described previously (16) and yeast transformations were by the method of Ito et al. (17).

RESULTS

We recently reported (16) that the expression of the UBI4 gene is induced by cAMP depletion through reduction of cAMP-dependent protein phosphorylation. Because this induction occurs in the absence of de novo protein synthesis, it seems likely that A-kinase directly regulates the activity of the transcriptional factor(s) by phosphorylation. Finley et al. (18) demonstrated that the UBI4 gene is a heat shock gene. From these results, we supposed that the heat response of the UBI4 gene may be mediated by the modulation of the intracellular cAMP level. However, experiments using mutants altered in the cAMP pathway clearly excluded this possibility (16). The UBI4 gene was clearly induced by heat shock in the bcy1 mutant, the ppd1 mutant, and the cyr1-1 mutant supplied with a sufficient amount of exogenously added cAMP. The hsp70 homolog, the SSA1 gene, was also induced by heat shock in the cyr1-1 mutant (see below).

Since heat shock induced UBI4 in the bcy1 mutant, we examined the acquiring of tolerance of the bcy1 mutant to a lethal temperature by heat shock.

TABLE 1
THERMAL TOLERANCE OF THE bcy1 MUTANT

Strain	% Survival		Fold stimulation
	Control	Heat shock	
RA1-13D	0.03	36.24	1208
RA1-13D-R	0.01	9.81	981

For determination, see METHODS.

As shown in TABLE 1, the degree of acquiring thermal tolerance is not different between wild type cells and bcy1 cells. However, the level of thermal tolerance of the bcy1 mutant is lower than that of wild type strain in both non-heat shocked and heat shocked conditions. Although, by a Northern hybridization experiment we could not detect any difference in the level of the steady state UBI4 mRNA between wild type and the bcy1 cells (FIGURE 1), the assay using the lacZ fusion genes indicate that the expression levels of the UBI4 and SSA1 genes are significantly lower in the bcy1 mutant (TABLE 2). These results are consistent with the fact that the bcy1 mutant is hyper-sensitive to acute heat shock (19).

Our data described above suggests that heat response is not mediated by cAMP, although the phosphorylation by A-kinase affects the expression level of the UBI4 gene. To reveal the mechanism of the dual regulation of the UBI4 gene by cAMP and heat shock in yeast, we dissected the promoter regions of the SSA1 and UBI4 genes. Although data is quite preliminary, the upstream activation

TABLE 2
STEADY STATE EXPRESSION LEVEL OF THE HSP GENES IN THE bcy1 MUTANT

Strain	Plasmid	β -galactosidase activity
RA1-13D	pJU26	51 ± 10
RA1-13D-R	pJU26	16 ± 1
RA1-13D	pJU6	103 ± 0
RA1-13D-R	pJU6	75 ± 3

Strains were cultured in SD medium supplemented with required amino acids. β -galactosidase activity was measured using exponential phase cells as enzyme sources.

TABLE 3
cAMP DEPLETION EFFECT IS NOT MEDIATED BY HEAT SHOCK ELEMENT

Plasmid	β -galactosidase activity		
	Control	Heat shock	-cAMP
pJU24	40 ± 5	66 ± 0	89 ± 5
pJU25	29 ± 4	71 ± 2	71 ± 1
pJU24-HSE	30 ± 3	30 ± 2	48 ± 4
pZJ-HSE2-26	180 ± 11	245 ± 8	75 ± 2
pZJ-HSE2-26R	158 ± 24	208 ± 17	52 ± 21

For determination, see METHODS.

sequence responding to cAMP levels seems to be different from the heat shock element. As shown in TABLE 3, the UBI4-lacZ fusion gene and the SSA1-lacZ fusion gene were induced by either heat shock or depletion of cAMP. There is a complete heat shock consensus sequence (11) in an upstream region of the UBI4 gene (12). AM18-5C-M1 carrying the UBI4-lacZ gene deleted in a heat shock box failed to induce β-galactosidase activity by heat shock, while cAMP depletion induced the activity. Slater and Craig constructed chimeric lacZ plasmids (pZJ-HSE2-26 and pZJ-HSE2-26R) in which the upstream activation site of the yeast cytochrome C gene, CYC1, was substituted by a single copy of heat shock elements of the SSA1 promoter plus its adjoining nucleotides (14). These chimeric plasmids responded to heat shock but not to cAMP depletion. These results suggest that cAMP response is mediated through another sequence different from heat shock elements. Identification of the cAMP responsive element is now underway on the SSA1 and UBI4 promoters.

Recently, it was shown that the UBI4 gene is induced in the stationary growth phase of the culture (18). Because the cAMP-pathway has an important role in the G0/G1 transition (2), we speculated that the induction of the UBI4 gene in the stationary phase of growth is mediated by the cAMP-pathway. To examine this hypothesis, we analyzed the time course of the level of the UBI4 mRNA in the strains RA1-13D and RA1-13D-R during exponential and stationary growth phases of the cultures (FIGURE 1). As expected, the UBI4 mRNA level increased during late log phase and was maintained in stationary phase in the wild type strain. On the other hand, in the bcy1 mutant, the UBI4 mRNA increased during late log phase but decreased to the basal level in the stationary phase of the culture. Thus, the induction of UBI4 in the stationary phase seems to be elicited by at least two distinct activation processes; one at late log and early stationary and the other during stationary phase. The cAMP-pathway seems to be involved in the latter. We speculate that the former activation process might be mediated through the heat shock element.

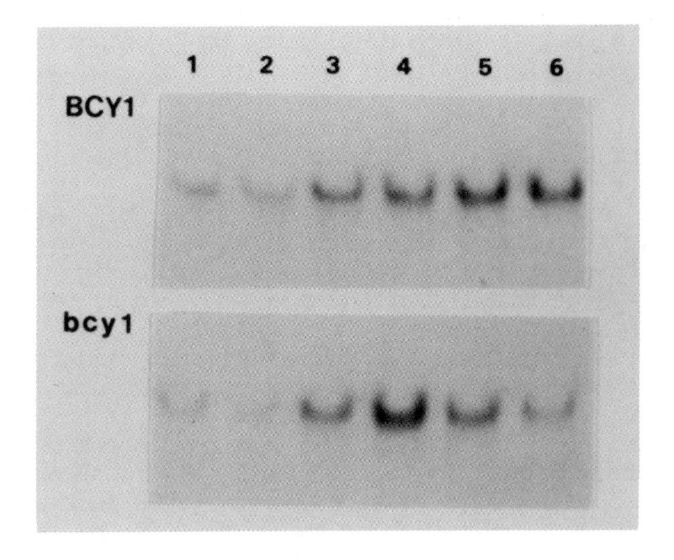

FIGURE 1. Time course of UBI4 expression during exponential and stationary growth phases of wild type and bcy1::URA3 cultures. 15 µg of total RNA from various growth phases of YPD cultures (lanes 1 and 2, exponential; lanes 3 and 4, late exponential; lane 5, early stationary; lane 6, stationary) were electrophoresed on a denaturing agarose gel. Northern hybridization was carried out using the UBI4 gene as a probe (16).

DISCUSSION

The fact that mutants altered in the cAMP-pathway are also altered in tolerance to a lethal temperature (19) suggested the involvement of the cAMP-pathway in acquiring thermal tolerance by heat shock. Shin *et al.* recently (19) showed that the bcy1-1 mutant was deficient in acquiring thermal tolerance by heat shock. However, we could not reproduce their results in the bcy1::URA3 mutant; that is, the bcy1::URA3 mutant normally acquired thermal tolerance by heat shock, although the level of thermal tolerance is lower than that of the wild type strain. We guess that the hyper-sensitive phenotype to acute thermal treatment in the bcy1 mutant can be explained by the slightly lower level of expression of heat shock genes in this mutant. Although the decrease of expression of individual genes is slight, the overall effect of many heat shock genes may have a drastic effect on the thermal tolerance of bcy1 cells.

We analyzed gene expression of the UBI4 and SSA1 genes in the mutants altered in the cAMP-pathway. Present work showed that cAMP regulates the expression of UBI4 and SSA1, but that induction of these genes by heat shock is not mediated through the cAMP-pathway. Dissection of promoters of UBI4 and SSA1 suggests that induction of heat shock genes by cAMP depletion occurs through upstream activation sequences different from heat shock elements. Our results suggest that heat shock genes in yeast are regulated by two completely divergent pathways, heat shock and cAMP depletion. However, the fact that the same genes are induced under these stresses indicate that there are, somehow, analogous biochemical natures in these stresses.

In wild type cells, the UBI4 gene was induced at early and late stationary phases of the culture. The latter induction seems to be mediated by the cAMP-pathway because the bcy1 mutant did not induce UBI4 at late stationary. It seems likely that the cAMP-pathway is required for maintenance of the high expression of heat shock genes during stationary phase or in a nutrient starvation condition.

ACKNOWLEDGMENTS

We thank J.Ishiyama and H.Mori of the Central Research Laboratory of Kikkoman Campany for their gift of cAMP.

REFERENCES

1. Iida H, Yahara I (1984). Durable synthesis of high molecular weight heat shock proteins in G0 cells of the yeast and other eukaryotes. J Cell Biol 99:199.

2. Matsumoto K, Uno I, Ishikawa T (1985). Genetic analysis of the role of cAMP in yeast. Yeast 1:15.

3. Özkaynak E, Finley D, Varshavsky A (1984). The yeast ubiquitin gene: head to tail repeats encoding a polyubiquitin precursor protein. Nature 312:663.
4. Margaret W-W, Stone DE, Craig EA (1987). Complex interactions among members of an essential subfamily of hsp70 genes in Saccharomyces cerevisiae. Mol Cell Biol 7:2568
5. Matsumoto K, Uno I, Ishikawa T (1984). Identification of the structural gene and nonsense alleles for adenylate cyclase in Saccharomyces cerevisiae. J Bacteriol 157:277.
6. Boeke JD, LaCroute F, Fink GR (1984). A positive selection for mutants lacking orotidine-5'-phosphate decarboxylase activity in yeast: 5-fluoro-orotic acid resistance. Mol Gen Genet 197:345.
7. Yamano S, Tanaka K, Matsumoto K, Toh-e A (1987). Mutant regulatory subunit of 3',5'-cAMP-dependent protein kinase of yeast Saccharomyces cerevisiae. Mol Gen Genet 210:413.
8. Maniatis T, Fritsch EF, Sambrook J (1982). "Molecular Cloning, a laboratory manual." Cold Spring Harbor Laboratory.
9. Stinchcomb DT, Mann C, Davis RW (1982). Centromeric DNA from Saccharomyces cerevisiae. J Mol Biol 158:157.
10. Botstein D, Falco SC, Stewart SE, Brennan M, Scherer S, Stinchcomb DT, Struhl K, Davis RW (1979). Sterile host yeasts (SHY): a eukaryotic system of biological containment for recombinant DNA experiments. Gene 8:17.
11. Pelham HRB (1982). A regulatory upstream promoter element in the Drosophila HSP70 heat-shock gene. Cell 30:517.
12. Özkaynak E, Finley D, Solomon MJ, Varshavsky A (1987). The yeast ubiquitin genes: a family of natural gene fusions. EMBO J 6:1429.
13. Ellwood MS, Craig EA (1984). Differential regulation of the 70K heat shock gene and related genes in Saccharomyces cerevisiae. Mol Cell Biol 4:1454.
14. Slater MR, Craig EA (1987) Transcriptional regulation of an hsp70 heat shock gene in the yeast Saccharomyces cerevisiae. Mol Cell Biol 7:1906.
15. Guarente L, Ptashne M (1981). Fusion of Escherichia coli lacZ to the cytochrome c gene of Saccharomyces cerevisiae. Proc. Natl. Acad. Sci. 78:2199.
16. Tanaka K, Matsumoto K, Toh-e A (1988). Dual regulation of the expression of the polyubiquitin gene by cAMP and heat shock in yeast. EMBO J 7:495.

17. Ito H, Fukuda Y, Murata K, Kimura A (1983). Transformation of intact yeast cells treated with alkali cations. J Bacteriol 53:163.

18. Finley D, Özkaynak E, Varshavsky A (1987). The yeast polyubiquitin gene is essential for resistance to high temperatures, starvation and other stresses. Cell 48:1035.

19. Shin D-Y, Matsumoto K, Iida H, Uno I, Ishikawa T (1987). Heat Shock response of <u>Saccharomyces cerevisiae</u> mutants altered in cyclic AMP-dependent protein phosphorylation. Mol Cell Biol 7:244.

Stress-Induced Proteins, pages 73–82

A STRUCTURAL UNIT OF HEAT SHOCK REGULATORY REGIONS[1]

John T. Lis, Hua Xiao, and Olga Perisic

Section of Biochemistry, Molecular and Cellular Biology, Cornell University
Ithaca, New York 14853

ABSTRACT Based on the observations summarized in this manuscript, we propose that the DNA sequences that specify heat shock regulation of adjacent genes are composed of contiguous arrays of a 5bp unit sequence (-GAA-) in alternating orientations. This proposal is supported both by in vivo assays of expression of hsp70 genes containing variant synthetic regulatory regions and by in vitro binding assays of heat shock factor to synthetic and native heat shock regulatory regions.

INTRODUCTION

Upstream regulatory regions of many eucaryotic genes contain a plethora of short sequence elements that each specify a site of protein binding (reviewed in 1). These sequences and the specific proteins with which they interact play central roles in setting the transcriptional level of genes. In some cases, a particular sequence element is present in multiple copies upstream of a gene as in the case of heat shock genes (reviewed in 2). In this manuscript, we will summarize work of others and some of our own attempts to define precisely both the DNA sequences required to create a heat shock inducible regulatory region and the

[1]This work was supported by a grant from NIH, GM25232.

interaction of these sequences with a specific protein, heat shock factor (HSF). In an accompanying manuscript (UCLA Symposium on Protein-DNA Interactions), we describe our approach involving protein-DNA crosslinking in vivo and nuclear transcription experiments that have led to the identification of a potential rate-limiting step in the activation of the major hsp70 gene and thus a potential target of catalytic action for the HSF-DNA complex.

The sequences required for the heat inducibility of the Drosophila hsp70 gene were first localized by Pelham (3) and Mirault et al. (4) by assaying a series of deletion mutants introduced into a monkey cell line, COS cells. A short region from nucleotide -66 to -47 upstream of the hsp70 gene was found to be essential for programming heat-induced expression in this system (3). Moreover, similarities between a sequence within this short region and sequences upstream of other heat shock genes were discovered (3), and a consensus CTGGAAT-TTCTAGA (later simplified to the 14 bp sequence C--GAA--TTC--G) was derived. This sequence, known as the heat shock element (HSE), was shown by Pelham and Bienz (5) to be capable of driving heat-induced expression in COS cells of a thymidine kinase gene to which it was fused. Regions containing good matches to the HSE were also shown to bind to a nuclear protein (6-8) that stimulates transcription of Drosophila heat shock genes in vitro (6) and in Xenopus oocytes (7).

Surprisingly, analyses of 5' deletion mutants of the Drosophila hsp70 gene in Drosophila cells (9) or germline transformants (10,11) revealed that deletions retaining the same promoter-proximal HSE showed only 1% of the normal heat-induced expression. The additional upstream sequences required to specify normal levels of induced hsp70 expression contain an additional match to the HSE (10). Thus, it was hypothesized that two copies of the HSE are required for high levels of heat shock-induced expression of the Drosophila hsp70 gene in Drosophila cells. As summarized here, the tests of this hypothesis by germline transformation have forced us to a new view of both the unit structure of HSEs and the interaction of these units with HSF (12).

RESULTS AND DISCUSSION

Activity of Hsp70 Genes Containing Variant Regulatory Regions

Hsp70-lacZ fusion genes containing variant heat shock regulatory regions were introduced into the Drosophila genome by P-element-mediated transformation (13). In this method, the DNA that is microinjected into embryos transposes with high efficiency into the genome. Stable lines containing single inserts of this fusion gene are then established and the expression of the inserted gene assayed. The fusion of the hsp70 gene to the E. coli-lacZ gene allows the expression of the introduced gene to be easily distinguished from that of the corresponding endogeneous hsp70 genes. Whole animal dissection assays in X-gal buffer permit the rapid assessment of the approximate level and tissue pattern of expression (14,15), while quantitative ß-galactosidase assays can be performed using the substrate, chlorophenol red ß-D-galactopyranoside, which yields a soluble product that absorbs at wavelengths distinct from the background absorbance of fly extracts (16).

An hsp70-lacZ fusion gene containing two naturally occurring HSEs (Sites I and II) of the hsp70 gene (-89 deletion shown in Fig.1a) is about 200-fold inducible in germline transformants, a level similar to that of the corresponding endogenous genes (12). This -89 deletion is defined here as the wild type to which the variant genes will be compared. Surprisingly, a synthetic regulatory region containing two perfect copies of the 14 bp heat shock consensus sequence shows only 7% of the heat-induced expression of the wild type (Fig. 1b). In this construct, the spacing between copies of the synthetic consensus sequence as well as the distance of these sequences to the TATA box sequnce is the same as in wild type. A notable difference between this synthetic regulatory region and wild type is in the sequences immediately flanking each consensus. Restoration of flanking sequence at the promoter-proximal end of Site I to the native hsp70 sequence (see underlined nucleotides of b) increases the level of expression to 27% (Fig. 1c). Results obtained with this and additional regulatory variants strongly implicate that sequences immediately flanking the 14 bp consensus have a role in the heat shock response (12). The role of sequences flanking HSEs has also been indicated from recent mutational analyses by Amin et al. (17).

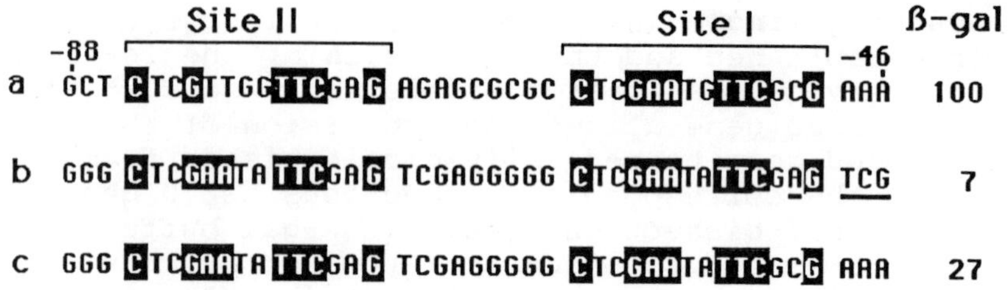

	-88	Site II		Site I	-46	ß-gal
a	GCT	CTCGTTGGTTCGAG	AGAGCGCGC	CTCGAATGTTCGCG	AAA	100
b	GGG	CTCGAATATTCGAG	TCGAGGGGG	CTCGAATATTCGAG	TCG	7
c	GGG	CTCGAATATTCGAG	TCGAGGGGG	CTCGAATATTCGCG	AAA	27

Figure 1. Sequences upstream of hsp70-lacZ genes whose activity has been assayed in Drosophila germline transformants. All constructs have the copies of the 14 bp consensus with the same spacing relative to each other and to the TATA box as the wild type gene. Heat shock-induced transformants containing single inserts were assayed and the average ß-galactosidase level (ß-gal) of several transformants of each variant is listed to the right of each sequence as the percentage of wild type activity (12).

As can be seen in Figure 1, the sequences that can partially restore the heat-induced activity (compare constructs b and c) recreate the sequence GAA at the end of Site I. This trinucleotide is also present twice in an inverted repeat symmetry (-GAA--TTC-) at the center of the 14 bp heat shock consensus (Fig. 2). Inspection of the genetically-defined regulatory regions of a variety of heat shock genes reveals that these 3 bases are the conserved central component of a 5 bp repeating

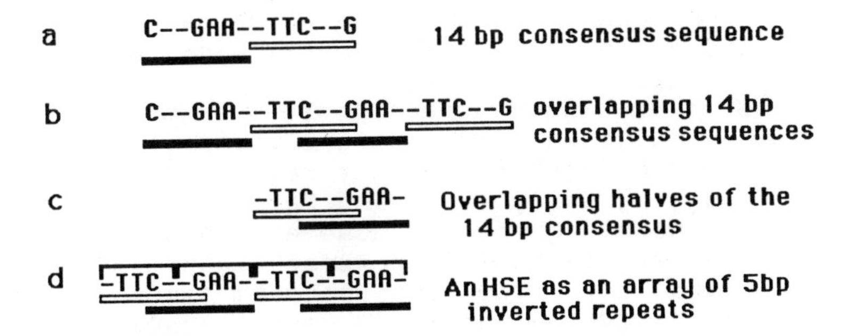

Figure 2. The relationship of the repeating 5 bp unit to the original 14 bp consensus. Regulatory regions of heat shock genes are known to possess multiple matches to the original 14 bp consensus that often overlap by 4 base pairs (reviewed in 2). When we searched regulatory regions of heat shock genes with each half of this consensus separately, we were surprised to find that the matches to right and left half consensus sequences lie in overlapping configurations more often than in the expected side-by-side configuration. This overlap creates a 10 bp sequence, which is itself a dyad of 5bp repeats (12). The array of four 5 bp repeats shown at the bottom of the figure encompasses the original 14 bp consensus sequence.

sequence. Site I and II of the hsp70 gene can each be regarded as an imperfect match to an array of four of these 5 bp repeats arranged in inverted orientations (or two direct 10 bp repeats) (12).

We tested the activity of an hsp70 gene variant in which the normal upstream regulatory region was replaced with two copies of the array of 5 bp repeats shown in Figure 2d. The two arrays in this variant were separated by 11 base pairs, a turn of the DNA helix. This gene is expressed at a 6-fold higher level than the hsp70 gene containing the pair of perfect 14 bp consensus sequences (construct b of Figure 1), and at a five-fold higher level than an hsp70 gene which has two native HSE elements separated by 11 base pairs (12).

Some heat shock genes possess arrays containing different numbers of inverted 5 bp repeats. The hsp83 regulatory region was identified originally as having three overlapping 14 bp consensus sequences; however, this region could be viewed instead to possess eight contiguous 5 bp repeats (12). The most upstream binding site for HSF (Site IV) in the hsp70 gene possesses a perfect match to three 5bp repeats. Thus, we propose that the description of an HSE can be simplified. The HSE can be veiwed as a repeating array of the 5 bp sequence -GAA- (or its complement -TTC-) where each repeat is inverted relative to the immediately adjacent repeats.

Within the 5 nucleotides in this basic -GAA- unit, the G is absolutely conserved and two As are highly conserved though some substitution for A with G may be permissible. Single base substitutions of the G or C base within this basic unit in derivatives of construct c of Figure 1 can reduce the heat-induced expression to background levels (12). A complete mutagenetic analysis of all five sites of the repeat to all possible alternative bases has not yet been performed.

A similar view of the heat shock regulatory region has been independently derived by R. Voellmy's group (18) using another set of regulatory region variants that have been assayed by transfection into cultured Drosophila cells. Moreover, their data indicates that a single 5 bp repeat can be skipped providing the 5 bp periodicity continues beyond the skipped sequence. Skipping 10 or 11 bp is less tolerated (12,18).

Binding of HSF to the 5 bp Units of the HSE.

The in vivo expression data described above demonstrate that sequences flanking the 14 bp consensus are important for heat shock regulation at Sites I and II of the hsp70 gene. High resolution ethylation interference mapping by Shuey and Parker (19) has shown that the three to four bases that extend beyond each side of these 14 bp consensus sequences are important for binding HSF. Thus, the region defined by these interference assays coincides well with the proposed larger 20 bp regulatory sequences for Sites I and II. If the 5 bp sequence is the basic unit recognized by HSF, then in the simplest case, the size of the footprint should grow by a 5 bp increment for each additional 5 bp repeat. Recent studies of binding of purified HSF to two variant heat shock regulatory regions that differ in number of 5 bp repeats suggest that this is the case. The footprints generated by DNAase I are precisely centered on a 15 bp sequence containing three 5 bp repeats, rather than on the 14 bp consensus that partially overlaps these three 5 bp repeats. Moreover, the footprint grows by precisely 5 bp when four repeats are used as a binding site (preliminary results). Also, methidiumpropyl-EDTA·Fe(II) footprints show that the regions of protection precisely match the beginning and end of such arrays. An analysis of binding of HSF to a complete series of regulatory regions that differ in the number of contiguous 5 bp repeats is in progress.

Each 5 bp unit could potentially be the target for binding of HSF or a domain of this protein. The interaction of a domain of protein with such a short sequence has precedent in the interaction of Zn fingers of the TFIIIA regulatory protein with the 5S gene of Xenopus (20). In the case of TFIIIA, there are multiple binding sites on a single polypeptide. In the case of Drosophila HSF, the number of binding domains per protein is not known at this time. If a single subunit of HSF binds to each 5 bp unit, then 8 monomers could interact with Sites I and II of the hsp70 gene. The size of the complex of protein and DNA of the heat shock regulatory region would be very large. Indeed, from electrophoretic analysis, Shuey and Parker (21) have estimated the size of the complex of HSF and a short DNA fragment containing Sites I and II to be a minimum of 500 kd. The stoichiometry of binding of HSF to heat shock regulatory regions will be important in understanding the molecular structure and dimensions of this regulatory complex.

Although we suggest that a 5bp sequence is a unit of interaction with HSF, we do not wish to imply that such a short domain provides for stable binding. In vitro studies of binding of purified HSF to variant regulatory regions indicate that a minimum of three adjacent 5bp repeats are required for generating a stable complex (unpublished results). The cooperativity of binding HSF to separated HSEs has been established (6). Such cooperativity may also play a role in binding HSF molecules to a repeating array of adjacent 5 bp units.

REFERENCES

1. Maniatis T, Goodbourn S, Fischer JA (1987) "Regulation of Inducible and Tissue-Specific Gene Expression." Science 236:1237.
2. Bienz M, Pelham HRB (1987) "Mechanisms of Heat-Shock Gene Activation in Higher Eukaryotes." Advances in Genetics 24:31.
3. Pelham HRB (1982) "A regualtory upstream promoter element in the Drosophila hsp70 Heat-Shock gene." Cell 30:517.
4. Mirault ME, Southgate R, Delwart E (1982) "Regulation of heat-shock genes: a DNA sequence upstream of Drosophila hsp70 genes is essential for their induction in monkey cells." EMBO J 1:1279.
5. Pelham HRB, Bienz M (1982) "A synthetic heat-shock promoter element confers heat inducibility on the herpes simplex virus thymidine kinase gene." EMBO J 1:1473.
6. Topol J, Ruden DM, Parker CS (1985) "Sequences required for in vitro transcriptional activation of a Drosophila hsp70 gene." Cell 42:527.
7. Wu C, Wilson S, Walker B, Dawid I, Paisley T, Zimarino V, Ueda H (1987) "Purification and Properties of Drosophila Heat Shock Activator Protein." Science 238:1247.
8. Sorger PK, Pelham HRB (1987) "Purification and characterization of heat-shock element binding protein from yeast." EMBO J 6:3035.

9. Amin J, Mestril R, Lawson R, Klapper H, Dreano M, Voellmy R (1985) "The heat shock consensus sequence is not sufficient for hap70 gene expression in Drosophila melanogaster." Mol & Cell Biol 5:197.
10. Dudler R, Travers AA (1984) "Upstream elements necessary for optimal function of the hsp70 promoter in transformed flies." Cell 38:391.
11. Simon JA, Sutton CA, Lobell RB, Glaser RL, Lis JT (1985) "Determinants of heat-shock-induced chromosome puffing." Cell 40:805.
12. Xiao H, Lis JT (1988) "Germline Transformation Used to Define Key Features of Heat-Shock Response Elements." Science 239:1139.
13. Rubin GM, Spradling AC (1982) "Genetic transformation of Drosophila with transposable element vectors." Science 218:348.
14. Lis JT, Simon JA, and Sutton CA (1983) "New heat-shock puffs and ß-galactosidase activity resulting from transformation of Drosophila with an hsp 70-lacZ hybrid gene." Cell 35:403.
15. Glaser RL, Wolfner MF, Lis JT (1986) "Spatial and temporal pattern of hsp26 expression during normal development." EMBO J 5:747.
16. Simon JA, Lis JT (1987) "A germline transformation analysis reveals flexibility in the organization of heat-shock consensus elements." Nucleic Acids Res 157:2971.
17. Amin J, Mestril R, Schiller P, Dreano M, Voellmy R (1987) "Organization of the Drosophila melanogaster hsp70 heat-shock regulation unit." Mol and Cell Biol 7:1055.
18. Amin J, Ananthan J, Voellmy R (1988) Mol & Cell Biol In Press.
19. Shuey DJ, Parker CS (1986) "Binding of drosophila heat-shock gene transcription factor to the hsp70 promoter: Evidence for symmetric and dynamic interactions." J Bio Chem 261:7934.
20. Fairall L, Rhodes D, Klug A (1986) "Mapping of the sites of protection on a 5SRNA gene by the Xenopus transcription - a model for the interaction." J Mol Biol 192:577.

21. Shuey DJ, Parker CS (1986) "Bending of promoter DNA on binding of heat-shock transcription factor." Nature 323:459.

Stress-Induced Proteins, pages 83–94

TRANSCRIPTIONAL REGULATION OF THE HUMAN HSP70 GENE[1]

Richard I. Morimoto, Dick Mosser, Terrill K. McClanahan, Nicholas G. Theodorakis, and Gregg Williams

Department of Biochemistry, Molecular Biology and Cell Biology, Northwestern University, Evanston, IL 60208

ABSTRACT We have examined promoter elements in the human HSP70 gene which regulate heat shock and metal induced stress, growth regulation and adenovirus E1A inducibility. A summary of results using mutant promoters and by reconstructing the HSP70 promoter onto the HSV-TK gene reveals that sequences between -68 to -64 containing the proximal CCAAT element are essential for wild type basal activity and E1A trans-activation. Induction of human heat shock gene transcription by heat shock and cadmium requires a single intact heat shock element at -107 of the HSP70 promoter. Both heat shock and cadmium induced transcription of the HSP70 gene occurs independent of protein synthesis. We have identified an HSE-binding factor from control cells which is easily distinguished from the stress-induced form of HSF which appears in heat shocked or chemically stressed cells proportional with promoter activity.

INTRODUCTION

Transcription of the human HSP70 gene is induced by conditions of stress such as heat shock, heavy metals and amino acid analogues, during normal cell growth by serum stimulation and transition through the cell cycle and in response to virus infection (1-6). The relationship between the growth activation of HSP70 gene expression and its induc-

[1]This work was supported by grants from the NIH, American Cancer Society, the March of Dimes Foundation and NSERC.

tion by viral encoded trans-activators is intriguing as is the recent evidence that HSP70 or HSP70-related proteins associate with other growth regulated proteins such as the cellular oncogene P53 (7,8).

The complex transcriptional regulation of the human HSP70 gene is mediated through multiple promoter elements, for example activation by heat shock and heavy metals utilizes distinct cis-acting sequences from those required for growth regulated expression (9-11). The sequences required for heat shock and metal ion induction map from -107 to -68 (9). Within these boundaries are two overlapping heat shock elements and a sequence analogous to the metal responsive elements of the metallothionein genes (12, 13). The sequences required for wild type levels of growth regulated expression are localized to the 5' flanking 68 nucleotides which corresponds to a CCAAT element, a purine-rich serum responsive element and the TATA element (9, 14, 15). Deletion of the CCAAT element from this promoter reduces both the overall levels of expression and the serum-inducible response.

Although it has been previously shown that HSP70 expression is also induced by metal ions and amino acid analogues, only for heat shock and dinitrophenol has it been demonstrated that the mechanism is through activation of heat shock factor followed by specific binding to the heat shock element (16-20).

RESULTS

Adenovirus E1A Induction of Chimeric HSP-CAT Genes: Analysis of 5'-Deletion Mutants

A schematic of the HSP70 promoter is shown in Figure 1. The effects of adenovirus E1A on transcriptional trans-activation of the HSP70 promoter were examined in transient assays of transfected HeLa cells containing the wild type or mutant promoters containing variable 5'-deletions fused to the CAT gene together with a plasmid that encodes E1A. In each series of HeLa cell transfections the following DNAs were introduced, the covalently linked HSP-CAT and HSP-NEO gene, the E3-CAT gene to act as a positive control for E1A trans-activation of an adenovirus promoter and pJOL a plasmid containing ad-5 E1A.

The results of a typical experiment reveal that

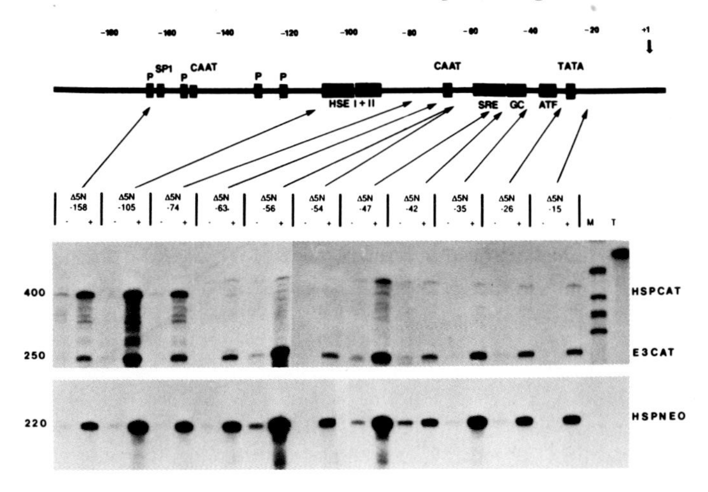

Figure 1. HSP70 Promoter Sequences Required for E1A Induction. HeLa cells were transfected with a mixture of Δ 5N (5' deletion mutants), the E3-CAT gene and either carrier plasmid (- lanes) or pJOL expressing E1A (+ lanes). Cytoplasmic RNA isolated 48 hrs. later was analyzed by RNAase protection assay. The 400 nuc. band corresponds to a HSP-CAT transcript and the 250 nt. band is from the E3-CAT gene. The HSP-NEO band of 220 nt is from protection by a separate NEO probe. M = molecular size markers, T = template probe.

adenovirus responsiveness requires sequences to -74 and decreases 10 to 20-fold in the deletion mutant extending to -63 (Figure 1). The sequences between -63 to -47 allow a very low level of E1A responsiveness while mutants extending to -47 and containing the TATA element here no longer E1A inducible. From these studies we conclude that the minimum sequences required for full E1A-inducibility extends from -74 to -63 corresponding to the CCAAT element in the basal promoter.

HSP70 Promoter Sequences that Confer Serum and Adenovirus E1A Responsiveness to the HSV-TK Gene

We independently assessed the role of specific HSP70 promoter elements by identifying sequences which could confer serum stimulation and E1A-inducibility on the thymidine kinase gene of Herpes Simplex Virus. The constructs shown in Figure 2 correspond to the wild type parental TK gene, the inactive TK gene which contains only the TATA element and the oligonucleotide promoter fusions. Serum stimulation was measured by the levels of chimeric RNA's in serum starved and stimulated cells. E1A-responsiveness was addressed by infection of stably transfected cell lines with wild-type adenovirus.

The addition of a 56 nucleotide fragment containing the CCAAT element and flanking purine-rich sequences to the -44 position of the TK gene generates a chimeric promoter in which the HSP70 CCAAT element is in a nearly identical position to the wild type HSP70 promoter. Pooled stable transfectants selected for neomycin resistance were serum deprived for 24 hrs and stimulated with 10% serum or infected with ad 5. While the levels of the pseudo-wild type TK mRNA were induced slightly by serum or infection, the level of the HSP-TK mRNA increased over 30-fold after 4 hrs of serum and 10-fold following infection. The serum stimulation kinetics for HSP-TK mRNA precedes endogenous HSP70 mRNA levels while the kinetics during infection are identical to that observed for the endogenous HSP70 mRNA (Figure 2).

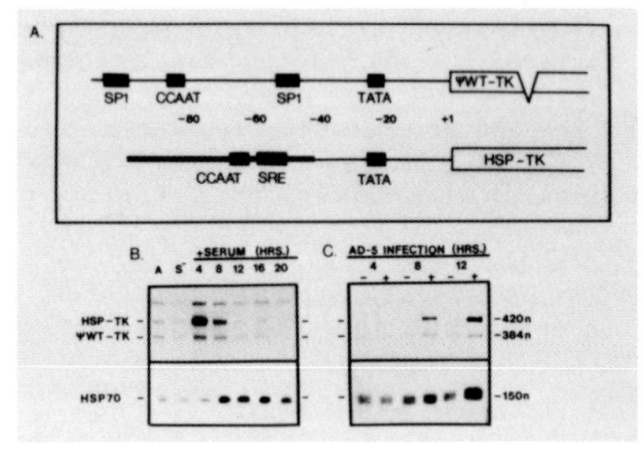

Figure 2. HSP70 Promoter Fusion to the HSV-Thymidine Kinase Gene. A. Organization of the wild type TK gene with upstream promoter elements and the HSP-TK gene which contains a 56 nucleotide fragment of -100 to -44 of the HSP70 promoter fused to the TATA of the TK gene. B. Stable pooled transfectants were serum deprived for 24 hrs. and stimulated (10% serum) for 4 to 20 hrs., RNA isolated and analyzed by RNAase protection. The 420 nt. band corresponds to the HSP-TK gene transcript and the 384 band is from the wtTK gene. The endogenous HSP70 transcript assayed by S1 protection yields a 150 nt. band. C. Time course of ad-5 infection. RNA was isolated and analyzed from mock of ad-5 infected cells. The bottom panel is the endogenous HSP70 transcript of 150 nt.

Metal Ions and Heat Shock Act Through a Single Heat Shock Element

Within a 39 nucleotide region from -107 to -68 are two overlapping HSE's corresponding to a distal 8/8 fit to the consensus and an overlapping proximal 4/8 fit to consensus and a sequence which corresponds to a metal responsive element from metallothionein genes. To determine whether both HSE's were required for heat shock induction and to directly identify the metal responsive element, we generated a series of linker-scanner and 5'-deletion mutations within this 39 nucleotide region. These promoter fusions were transfected into HeLa cells, the cells were either subjected to heat shock or incubation with cadmium and the levels of HSP-CAT mRNA measured by nuclease protection of a uniformly labeled riboprobe template.

The mutant promoter elements are diagramatically shown in Figure 3. The upstream boundary of the HSE is at -107, mutants LSN 107/100 and delta 5N-105 in which either a consensus nucleotide or non-conserved nucleotide are replaced, exhibit levels of heat shock and cadmium induction equal to that observed for the wild type LSN promoter. The upstream boundary can be accurately positioned by the location of mutations in the LSPN promoter which substitute the consensus nucleotides between -107 to -102. Transfection of LSPN-CAT results in a promoter void of both heat shock and cadmium responsiveness. These results indicate that the distal highly conserved 8/8 fit to the consensus HSE is essential for both heat shock and metal ion induction transcription.

The downstream boundary can be accurately positioned from the mutant LSN 94/84 which changes the consensus G at position -94 to a C and further reduces the weak 4/8 fit in the proximal HSE to a 3/8 fit to consensus. These changes in LSN 94/84 result in a promoter which retains both heat shock and cadmium responsiveness even though only a single HSE remains.

Kinetics of HSP70 Gene Transcription by Heat Shock and Heavy Metal Ions: Effects of Protein Synthesis Inhibitors

The kinetics of HSP70 gene transcription following heat shock or cadmium treatment were examined in cells maintained at 42°C or in the presence of the metals for periods up to 6 hours. Transcription rates were measured in isolated nuclei

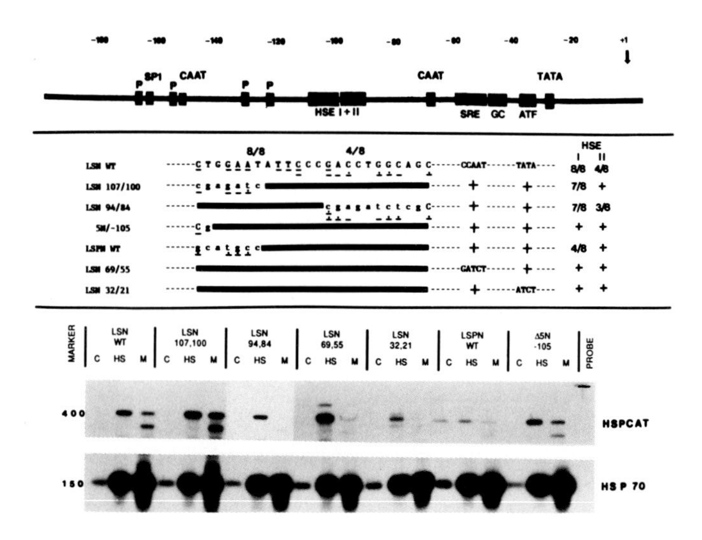

Figure 3. Sequences Required for Heat Shock and Cadmium Induction. The sequences in two overlapping HSE's located by -207 to -83 were mutated by 5' deletion or linker-scanner replacements. The consensus nucleotides in the wt promoter are underlines, mismatch with consensus are marked with a dot. Mutated sequences are noted by the long black bar; the effects on the two HSE's are summarized relative to the 8/8 fits for HSE I and 4/8 fit for HSE II (+ = wt). The bottom panel corresponds to RNAase protections for HSP-CAT RNAs and endogenous HSP70 mRNA in control, heat shock and cadmium treated cells.

using in vitro transcription in the presence of ^{32}P-UTP, the nascent radiolabeled RNA's were isolated and allowed to hybridize to cloned gene probes bound to nitrocellulose. The results were quantitated by scanning densitometry (Figure 4).

During continuous heat shock the rate of HSP70 gene transcription rapidly increases to maximal levels by 30 min. and declines to near background levels by 120 minutes.

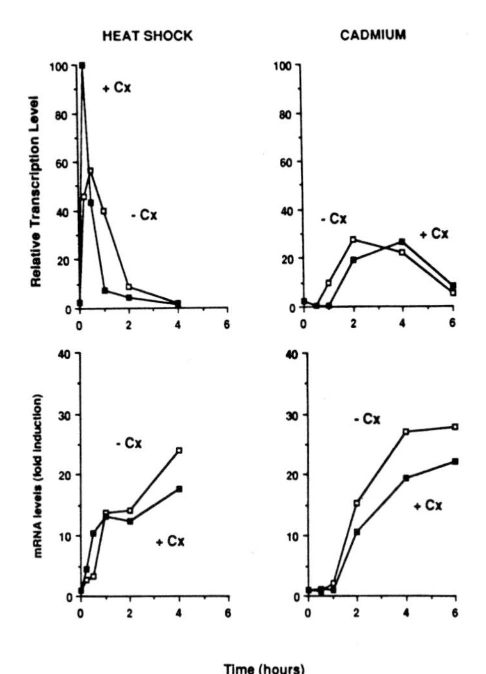

Figure 4. Transcription Rate and mRNA Levels in Heat Shocked and Cadmium Treated Cells-Requirement for Protein Synthesis. HeLa cells were continuously heat shocked (42C) in the presence or absence of cycloheximide (Cx) or treated with cadmium. Transcription rates were measured by nuclei runoff analysis and mRNA levels measured by S1 protection. The results were quantified by scanning densitometry.

In contrast, cadmium treatment results in a delayed response, requiring approximately 60 min. before transcription rates increase with the levels persisting through 6 hours of incubation.

We examined the effects of protein synthesis inhibitors on induction of HSP70 gene transcription to determine whether the tightly regulated expression observed in the previous experiments was dependent on the presence of newly synthesized proteins. A duplicate experiment performed in the presence of 100 µg/ml cycloheximide, corresponding to a level sufficient to block 99% of protein synthesis including HSP70 synthesis, reveals that the overall kinetics of induction and repression are similar in the presence or absence of protein synthesis. However, in the absence of protein synthesis the rate of HSP70 gene transcription is 40% higher at the 30 min. time point and declines rapidly to background levels by 60 min. The effect of protein synthesis inhibitors on cadmium induction results in a 60 min. delay in transcription induction but a similar pattern thereafter.

Kinetics of HSF-Binding Activity Following Heat Shock and Metal Ions

Control (37C) HeLa cells contain a HSE binding activity which is competed by the homologous oligonucleotide but not by heterologous oligonucleotides (Figure 5). The mobility of the control HSE binding activity is distinct from the slower migrating gel shift observed using extracts from heat shocked cells. The dissociation half-life is 150 min. for the HSF:HSE complex from control cells and 40 min. for the stress-induced complex. The direct DNA contacts with the control or stress-induced forms of HSF were identified using methylation interference. Interestingly, we find that the patterns are similar but not identical suggesting subtle differences in how the control or stress-induced forms of HSF recognize HSE (11).

In experiments parallel to Figure 4, cells were heat shocked at 42C or incubated with 30uM Cadmium sulfate for periods up to 6 hours. Plates of cells were taken at various time points to examine the level of HSF by gel shift assays. Near maximal levels of the induced form of HSF appear after 15 min. at 42C and begin to decline between 120 to 240 min. (Figure 6). In the presence of cycloheximide, HSF is still rapidly activated to near maximal levels at 15 min., however the levels of HSF decline rapidly.

Extracts from cadmium treated cells contain high levels of an HSE-binding activity which appears between 30-60 min., reaches maximal levels between 120-240 min. and declines by 360 min. The relative levels of HSF shown in Figure 6 corresponds well with the absolute rate of cadmium-induced transcription. By mobility on non-denaturing polyacrylamide gels, the metal-induced HSE-binding activity is indistinguishable from the heat-induced factor. In the presence of cycloheximide, the levels of HSF are delayed approximately 60 min., reaches maximal levels between 120-240 min. and declines by 360 min.

DISCUSSION

We have examined the role of 5'-flanking sequences in the human HSP70 promoter which are required for basal expression, transcriptional transactivation by adenovirus E1A heat shock and cadmium induction. The compact organization of promoter elements upstream of the HSP70 gene offers a system to dissect and understand how multiple cis-acting transcription elements are utilized to confer specific forms of cellular and viral inducible regulation.

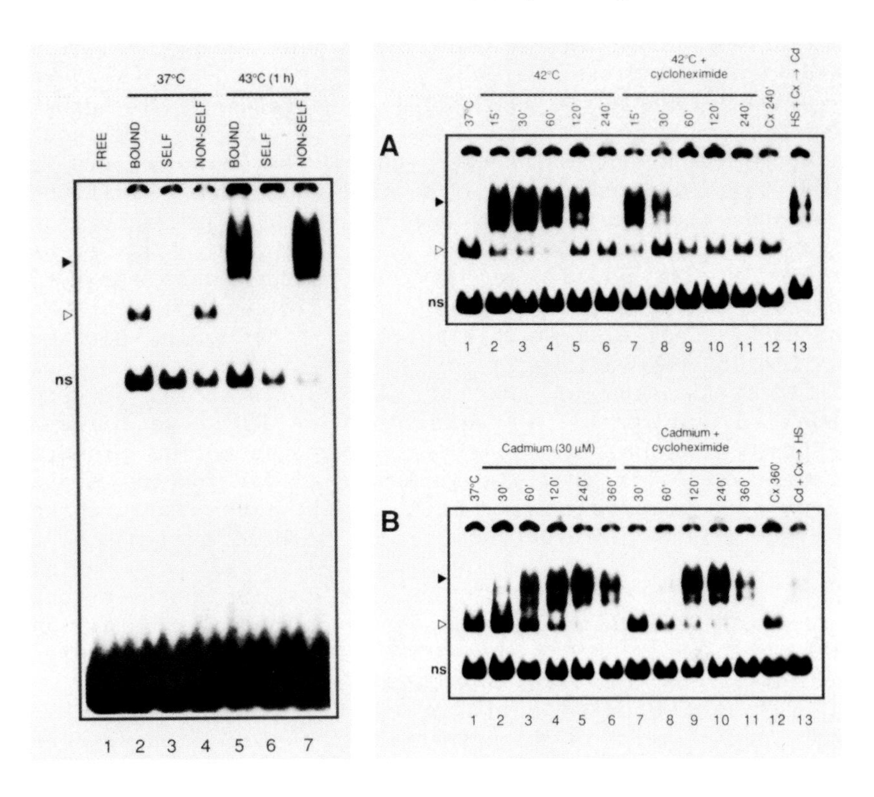

Figure 5. Left Panel. HSE-Binding Activities from Control and Heat Shocked HeLa cells. A synthetic oligonucleotide corresponding to -107 to -83 of the HSP70 promoter was used for gel mobility shift assay. Binding in the absence of protein (lane 1), presence of control (lane 2) or heat shock (lane 5) extract, competition with excess cold HSE (lanes 3, 6), competition with excess cold oligonucleotide containing the CCAAT element (lanes 4, 7). Open arrow-control HSF, filled arrow-stress-induced HSf, ns-non-specific band.

Figure 6. Right Panel. Kinetics of Stress-Induced HSF During Heat Shock (A) and Cadmium (B) Treatment. Whole cells extracts were prepared from duplicate aliquots of samples in Figure 4 and used for gel shift analysis.

Activation of the cellular HSP70 gene during adenovirus infection is mediated through the E1A-13S product (5). The promoter elements of the HSP70 gene required for growth regulation and E1A trans-activation may be the same. Although the sequences between -68 and -64 are necessary for full E1A responsiveness, it is clear that additional sequences must be necessary because the CCAAT element is an essential upstream element for many eukaryotic genes transcribed by RNA polymerase II. While our results clearly indicate that the CCAAT element is sufficient it may be that the HSP70 CCAAT element is atypical of other genes which are not E1A-13S responsive.

Identification of the HSE as the cis-acting sequence required for metal ion activation of HSP70 gene transcription demonstrates that a single cis-acting promoter element mediates multiple forms of stress-induced transcriptional activation. Heat shock gene transcription through HSF-HSE interactions can be induced by temperature elevation, metal ions, amino acid analogues, arsenite and DNP (11, 16, 19). However, the kinetics of transcriptional activation differs greatly between heat shock and metal ions with heat shock having immediate effects (within minutes) and metal ions requiring approximately 60 minutes.

The HSE binding activity identified in control HeLa cells has distinct electrophoretic characteristics from the HSE binding activity from heat shocked and metal ion treated cells. Although the apparent mobility of the DNA-protein complex from control or stressed cells are distinct on native gels, the patterns of DNA:protein contacts are similar but not identical (11). Our results would suggest either that two forms of HSF exist in the cell and that these two forms could convert in vivo or the second possibility that HeLa cells have multiple HSE-binding proteins.

Comparison of the levels of HSF to the rates of transcription reveals a close correspondence between the absolute in vivo rate of heat shock induced transcription with the levels of the HSE-binding factor. The decline of HSP70 gene transcription during continuous heat shock in the absence of protein synthesis would appear to be due to the lack of stress-induced HSF. Our results also reveal that protein synthesis is neither required for the activation of HSP70 gene transcription nor for the appearance of the heat shock induced form of HSF, however, protein synthesis is required to maintain the continuous high levels of both HSF and HSP70 gene transcription rates.

It is clear from these results and previously from other laboratories (11, 16, 19) that HSF is not de novo synthesized and must be present in a transcriptionally inert state. We have identified a control form of HSF and a stress-induced form (heat shock, metal ions and amino acid analogues) of HSF which binds to the human HSE with sequence specificity. If both control and activated forms of HSF exist in vivo, how is transcription regulated? One explanation for yeast HSF which appears to be bound in control cells is that heat shock stimulates post-translational modifications of HSF thus resulting in a transcriptionally activated form (17). Alternatively, the affinity of the control and activated forms of HSF to HSE could differ. It appears that the inherent thermostability of HSF-HSE complexes exhibited by stress-induced HSF relative to control HSF (11) might offer a simple explanation for both forms of HSF and how transcriptional regulation by heat shock is ensured.

REFERENCES

1. **Wu BJ, Hunt C, and Morimoto R** (1985). Structure and expression of the human gene encoding major heat shock protein HSP70. Mol Cell Biol **5**:330.
2. **Wu, BJ and Morimoto R** (1985). Transcription of the human HSP70 gene is induced by serum stimulation. Proc Natl Acad Sci USA **82**:6070-6074.
3. **Milarski K, and Morimoto R** (1986). Expression of human HSP70 during the synthetic phase of the cell cycle. Proc Natl Acad Sci **83**:9517.
4. **Hickey E, and Weber L** (1982). Modulation of heat shock polypeptide synthesis in HeLa cells during hyperthermia and recovery. Biochem **21**:1513.
5. **Wu B, Hurst H, Jones N, and Morimoto R** (1986). The E1A 13S product of adenovirus 5 activates transcription of the cellular human HSP70 gene. Mol Cell Biol **6**:2994.
6. **Nevins JR** (1982). Induction of the synthesis of a 70,000 dalton mammalian heat shock protein by the adenovirus E1A gene product. Cell **29**:913-919.
7. **Pinhasi-Kimhi O, Michalovitz D, Ben-Zeev A, and Oren M** (1986). Specific interaction between the P53 cellular tumour antigens and major heat shock proteins. Nature **320**:182.
8. **Hinds P, Finlay C, Frey A and Levine A** (1987). Immunological evidence for the association of p53 with a heat shock protein, hsc 70, in 053-plus-ras transformed cell lines. Mol Cell Biol **7**:2863.
9. **Wu BJ, Kingston RE, and Morimoto RI** (1986). Human HSP70

promoter contains at least two distinct regulatory domains. Proc Nat Acad Sci USA **83**:629-633.

10. **Williams G, McClanahan T, and Morimoto R**. Identification of sequence in the human HSP70 promoter required for adenovirus EIA responsiveness (in preparation).
11. **Mosser D, Theodorakis N, Williams G and Morimoto R**. Induction of human HSP70 gene transcription by heat shock, metal ions and amino acid analogues: Promoter elements and factor requirements (in preparation).
12. **Karin M, Haslinger A, Holtgreve H, Richards RI, Krauter, P, Westphal HM and Beato M** (1984). Characterization of DNA sequences through which cadmium and glucocorticoid hormones induce human metallothionienin-IIa gene. Nature (London) **308**:513-519.
13. **Stuart GW, Searle PF, Chen HY, Brinster RL and Palmiter RD** (1984). A 12-base-pair DNA motif that is repeated several times in metallothioienin gene promoters confers metal regulation to a heterologous gene. Proc Natl Acad Sci (USA) **81**:7318-7322.
14. **Morgan WD, Williams GT, Morimoto RI, Greene J, Kingston RE, and Tjian R** (1987). Two transcriptional activators. CCAAT-box binding transcription factor and heat shock transcription factor, interact with a human HSP70 gene promoter. Mol Cell Biol **7**:1129-1138.
15. **Greene JM, Larin Z, Taylor ICA, Prentice H, Gwinn KA, and Kingston RE** (1987). Multiple basal elements of a human HSP70 promoter function differently in human and rodent cell lines. Mol Cell Biol **7**:1129-1138.
16. **Kingston RE, Schuetz, TJ and Larin Z** (1987). Heat-inducible human factor that binds to a human HSP70 promoter. Mol Cell Biol **7**:1530-1534.
17. **Sorger PK, Lewis MJ, and Pelham HRB** (1987). Heat shock factor is regulated differently in yeast and HeLa cells Nature (London) **329**:81-84.
18. **Wu C, Wilson S, Walker B, Dawid I, Paisley T, Zimarino V, and Ueda H** (1987). Purification and properties of _Drosphila_ heat shock activator protein. Science **238**:1247-1253.
19. **Zimarino V, and Wu C** (1987). Induction of sequence-specific binding of _Drosophila_ heat shock activator protein without protein synthesis. Nature (London) **327**:727-730.
20. **Wiederrecht G, Shuey D, Kibbe W, and Parker C** (1987). The Saccharomyces and Drosophila heat shock transcription factors are identical in size and DNA binding properties Cell **48**:507.

Stress-Induced Proteins, pages 95–104

GENETIC ANALYSIS OF THE HEAT SHOCK RESPONSE IN *DROSOPHILA* [1]

J. José Bonner, Mark Hallett, Monica McAndrews, and Janice Parker-Thornburg

Institute for Molecular and Cellular Biology and Department of Biology, Indiana University, Bloomington, IN 47405

ABSTRACT We have sought means of identifying the genes involved in the regulation of the heat shock response in *Drosophila melanogaster*. One approach is to generate a gene fusion that links the structural gene for Alcohol Dehydrogenase (ADH) to the hsp70 promoter, and create transgenic flies in which ADH is heat shock inducible. Using selection for ADH activity, we have isolated mutations that result in the constitutive expression of heat shock genes. Using selection against ADH, we have sought mutations that prevent the expression of heat shock genes during stress. These screens have met with little success; perhaps such mutations are deleterious. A second approach is to examine the embryonic phenotypes of flies homozygous for chromosomal deletions, to map the locations of genes that are essential for the induction of heat shock genes during stress. This method does not require that mutations in the genes be viable. Using this approach, we have identified several loci which appear to be required for a normal response to heat shock.

INTRODUCTION

The transcriptional activation of heat shock genes, such as hsp70, depends upon the interaction of a Heat Shock Factor protein (HSF) with a *cis* -acting Heat Shock Element (HSE) (1,2). In *Drosophila* and human cells, HSF binds the HSE only during stress. Because the DNA-binding activity of HSF can be induced even in the presence of inhibitors of protein synthesis (3), it appears that HSF exists in an inactive state in normal cells, and is activated in response to stress. The activation mechanism in *Saccharcmyces*

[1]This work was supported by grant GM26693 from the National Institues of Health

cerevisiae appears to include phosphorylation (4), although additional modifications cannot be ruled out.

Relatively little is known of the mechanism whereby HSF is activated on heat shock. Less is known of the enzyme(s) which catalyzes the activation. It, too, must be activated during stress (3), but whether through enzymatic or allosteric means is unknown.

While there would thus appear to be at least a two-step cascade of regulatory activities involved in the activation of transcription of heat shock genes, there are as yet no clues whether the factors "upstream" of HSF are dedicated solely to the heat shock system, or whether they may have other physiological functions as well. One can easily imagine that the kinase suggested by the work of Sorger and Pelham (4) would have additional substrates besides HSF itself.

In the face of such complexity, it would seem valuable to have genetic evidence that could help answer some of the relevant questions. To this end, we have taken two approaches toward the isolation of mutations which affect the regulatory system. One has been the directed search for heat shock mutants using an hsp70-ADH gene fusion as a selectable marker. The other has been a more general screen for loci that must be present for a normal response to occur.

RESULTS

Dominant Mutations

In order to select mutations that affect the heat shock system, we generated transgenic flies carrying an hsp70-ADH fusion gene (5); the endogenous ADH gene was nonfunctional. The hsp70-ADH fusion gene provides a selectable marker: flies that are ADH^+ (ie, after heat shock) survive exposure to ethanol, and die on exposure to 1-pentyne-3-ol (pentynol), while flies that are ADH^- (non-shocked) are killed by ethanol and survive pentynol.

The phenotype expected of flies that cannot respond to heat shock with the induction of the heat shock proteins (hsps) is not certain. On the basis of the implication of the hsps in the development of thermotolerance, it is conceivable that non-responding mutants would die during the shock. To recover mutants as *viable* flies, we sought instead to isolate mutations that confer a *constitutive heat shock* (*chs*) phenotype. These should be isolable as flies that survive exposure to ethanol without a prior heat shock.

From among 33,000 flies screened in the F1 generation after mutagenesis with EMS, we recovered some 21 mutations

(6). Several of these we examined in detail. They had the following characteristics: they (a) act in *trans* on the hsp70-ADH fusion, (b) activate an hsp26-ADH fusion, and (c) fail to activate an hsp70-ADH fusion carrying a deletion of the HSE. We interpret these observations to mean that the mutations act to induce the heat shock response in the mutant flies, rather than to activate the hsp70-ADH fusion in some other way.

Upon heat shock of a wild type fly carrying the hsp70-ADH fusion, virtually all cell types accumulate the enzyme, as assayed by histochemical staining (5). Therefore, mutations in factors intimately involved in the heat shock response should be expected to affect all cell types. Interestingly, the *chs* mutations appear to affect only some tissues. This is shown in figure 1, as frozen sections of several *chs* mutants stained for *constitutive* ADH activity, and summarized in figure 2.

Although the *chs* mutants exhibit tissue-specific staining for ADH activity under normal growth conditions, those which we have examined exhibit ubiquitous staining after heat shock. This suggests that the heat shock system

figure 1

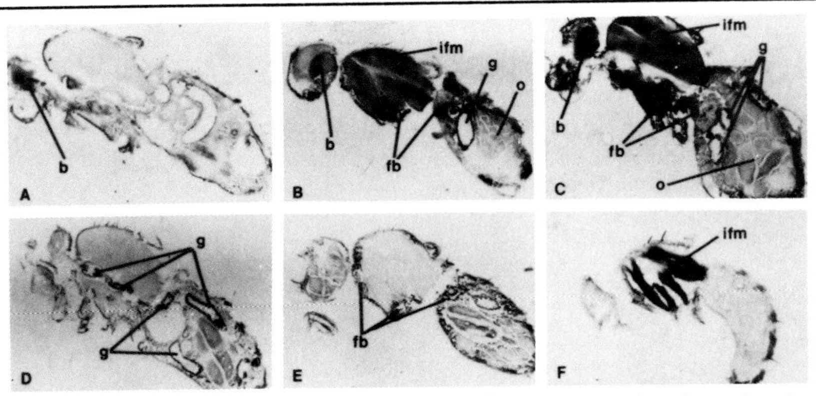

ADH activity in several *constitutive heat shock* mutants. Flies (*not* heat shocked) were frozen and sectioned, and the sections stained for ADH activity (6). A: ADH-deficient fly, stained for 24hrs; all others were stained ≤ 2hr. B: hs-ADH, shocked twice daily at 37°. C: mutant u144. D: mutant g65. E: mutant 181. F: mutant $Act88F^{KM75}$. B-F: all carry the Adh^{fn23} null allele of Adh, the hs-ADH fusion, and the mutation noted. b = brain, ifm = indirect flight muscle, fb = fat body, g = gut, o = ovary.

Figure 2

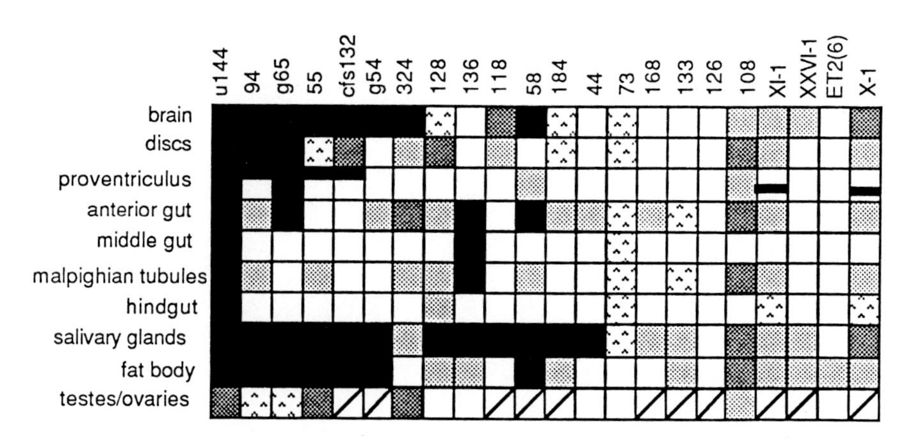

ADH activity in various mutants. Larvae were dissected and stained for ADH activity (without heat shock). The boxes are filled roughly in proportion to the intensity of staining of the different tissues, with the exception of mutants XI-1 and X-1 proventriculus, for which staining was observed in a discrete ring surrounding the tissue.

is intact in these mutants, and that the mutations act to trigger the heat shock response in a cell-type-specific manner. This suggestion is confirmed by the observation that the mutation, *Act88F*^{KM75}, in the actin gene expressed only in the indirect flight muscle (IFM), induces the heat shock system in the IFM only. We therefore suggest that these *chs* mutations represent lesions in genes that are expressed in a tissue-specific manner, and that the reason they induce the heat shock response is that they generate an internal "biochemical stress" in the cells that express the mutation.

What is the nature of the biochemical stress? One possibility could be that the protein products of the mutant genes are recognized as aberrant by the cells in which they are expressed. According to this model, the aberrant protein would itself trigger the heat shock response. We consider this model unlikely for two reasons. First, we have been unable to reconstruct a *constitutive heat shock* phenotype by crossing a variety of mutant genes into the background of our hsp70-ADH fusion (5). Second, we note that there are very few alleles of Act88F that induce the heat shock response (7)(8). Those that do carry mutations in the carboxy-terminal 10% of the protein, and generate a relatively stable protein product. Those that do not induce the heat shock system generate a protein product that is rapidly degraded. Since the latter are clearly recognized by the cell as aberrant protein, and are thus degraded, it appears unlikely that the aberrant nature of the protein *per se* is what triggers the heat shock response.

It is more likely, we feel, that the mutations we have induced, and *Act88F*^{KM75} as well, interfere with cellular physiology by virtue of their inherent biochemical activities. In doing so, they perturb what is normally a delicately balanced system, resulting in rather profound and pleiotropic alterations in several different biochemical systems. According to this model, the heat shock response would be induced by an indirect effect on the system's normal trigger.

Recessive Mutations

The *chs* mutations provide a unique situation: the heat shock system is induced without the need to heat shock the fly. Potentially, this offers a means of circumventing the caveat raised above, that non-responding mutants might succumb to the very heat shock that is required to induce the hsp70-ADH fusion. To test this possibility, we mutagenized flies of genotype *+/Y ; Adh*^{fn6}*cn ; hs-ADH Act88F*^{KM75}, crossed them to flies of genotype *C(1)RM ; Adh*^{fn6}*cn ; hs-ADH Act88F*^{KM75}, and selected the F1 progeny with pentynol. This protocol should recover recessive X-linked mutations (or dominant mutations on any chromosome) that act as suppressors of the ADH^{+} phenotype of these flies.

From some 50,000 X chromosomes screened in this fashion, no valid suppressors were recovered. This represents a mutation rate of $<2x10^{-5}$, which is very low for EMS-induced mutations. We entertain two possibilities for the failure to recover mutations on the X chromosome: (a) there may be no genes required for the heat shock response on the X chromosome, or (b) such mutations may be lethal.

It is at present difficult to predict the whole-organism phenotype of a mutation in HSF, or in one of the factors involved in its activation. *If* the hsps are required for viability at normal growth temperatures, and *if* their basal-level expression depends on HSF in its active mode, then mutations in HSF or its activators will be lethal. *If* the activators of HSF are not dedicated solely to the heat shock response, but have other substrates as well, and *if* those other substrates play essential physiological roles, then, again, mutations will be lethal. These assumptions are not unreasonable; indeed, the former have some support (9). Lethality of mutants which cannot respond to heat shock would appear to make the genetic analysis of the system somewhat less straightforward.

Brute Force

By definition, lethal mutations of *Drosophila* fail to reach adulthood. However, many die late in development, and very, very few die before cellular blastoderm. There is thus a window in *Drosophila* development between blastoderm, when the heat shock response first becomes detectable (10) and the effective lethal phase of any given mutation, during which time a mutant carrying that mutation can be examined for its ability to mount a normal heat shock response.

We have therefore begun a systematic examination of the *Drosophila* genome for loci which are essential for the induction of hsp synthesis. We have collected embryos from appropriate fly stocks, heat shocked them midway through embryonic development, then labeled *individual* embryos with ^{35}S-methionine, and displayed the proteins of the various individuals on SDS polyacrylamide gels. We have used two general kinds of fly stocks: deletions, and compound chromosomes. The former generate embryos deficient for "small" regions of the genome, while the latter generate embryos deficient for entire chromosome arms.

figure 3

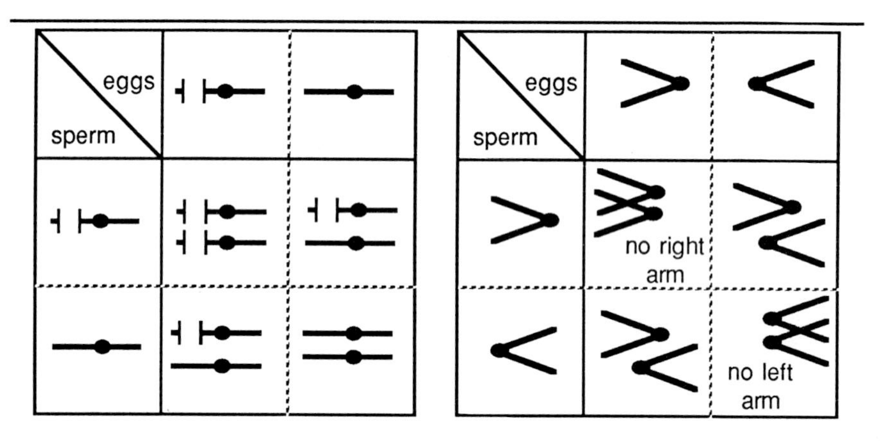

Punnet squares for progeny derived from self-cross of deletion heterozygote (left) and compound autosomes (right), illustrating the recovery in 25% of progeny of either deletion homozygotes, or complete deficiency of entire chromosome arms.

figure 4

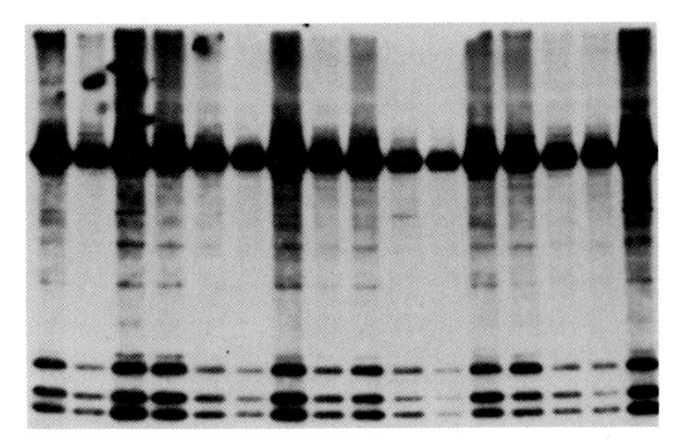

Heat shock response of individual embryos from a stock of *C(1)M4 /0 x X^Y /0* . Embryos were collected, aged 6-9hrs, shocked at 37° for 1hr, dechorionated and cut in half individually in 3μl drops of TB1 (5) containing ^{35}S-methionine. After labeling for 1hr, proteins were acetone-precipitated and run on an SDS polyacrylamide gel.

Surprisingly, removal of the entire X chromosome (using *C(1)M4 /0 x X^Y /0)* had no effect on the heat shock response (figure 4). Similarly, removal of the right arm of the second chromosome had no effect (not shown).

That embryos lacking these chromosomes gave a normal heat shock response can be explained in several ways. First, it is there may be no genes on these chromosome arms that are required for the response. For the X chromosome, this would be consistent with the failure to recover mutations as suppressors of the *Act88F^{KM75}* mutation. It is also conceivable, however, that there *are* relevant genes on these chromosomes, but that their products are packaged into the egg during oogenesis in great enough quantities to sustain the embryos beyond the time at which we performed the experiment. We cannot distinguish between these two possibilities.

Chromosome arms 2L, 3L, and 3R were more interesting. For all three, we have found at least one deletion stock that, in our initial tests, appears to give non-responding embryos. One of these is shown in figure 5, and a summary of results to date is shown in figure 6.

figure 5

Heat shock response of individual embryos from a stock of *Df(2L)ed*^{dph1}*/Cy*. Embryos were collected, aged 12-16 hrs, and treated as for figure 4.

figure 6

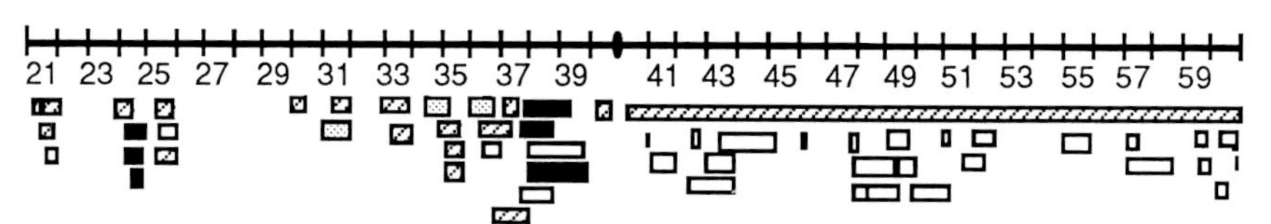

Chromosome 3

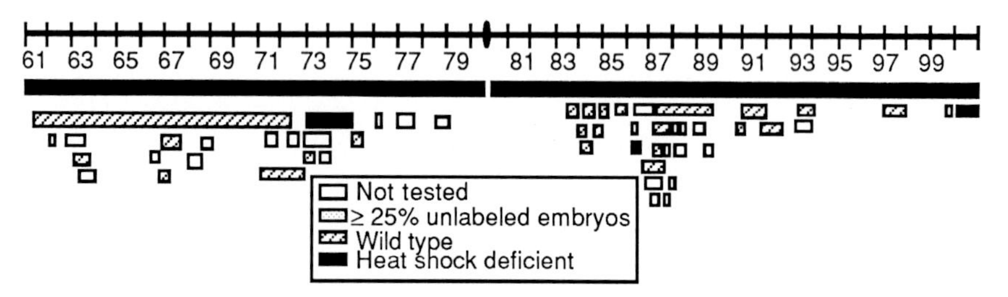

Summary of autosomal deletions tested for effects on the heat shock response.

Although it appears likely that the non-responding embryos in experiments like that of figure 5 are homozygous for the deletion, we have as yet no direct proof that this is the case. For most of the deletions we have studied, the homozygotes die with no morphological abnormalities that would allow us to identify them unambiguously. Crosses are in progress to recombine mutations in segmentation genes (which provide an embryonic morphological marker) onto chromosomes bearing the most interesting of the deletions, but until these crosses are complete, the conclusions from these experiments must remain tentative.

Two features of the data are noteworthy. First, the frequency of non-responding mutants deviates from the expected Mendelian ratio of 25% (ranging from 15% to 39%). This probably reflects the relatively small sample sizes we have been able to handle in these experiments, but may reflect a more significant biological factor. Second, many genotypes, like that of figure 5, exhibit embryos with "intermediate" hsp profiles: absence of hsps 22, 23, 26, and 28, but moderate synthesis of hsp70. If this is a partial response due to incomplete activation of HSF, it differs significantly from the partial response seen in wild type flies at submaximal heat shock temperatures (eg, 34°), in which synthesis of hsps 22, 23, 26, and 28 is generally greater than synthesis of hsp70.

Thus far, 5 autosomal loci have tested positive (albeit with the caveat mentioned above) in this screen for loci required for a normal heat shock response. This is probably an underestimate, for two reasons. First, we have not examined all of the genome; deletions are not available for all chromosomal regions. Second, this approach will fail to detect those loci for which significant quantities of gene product are loaded maternally during oogenesis.

Nonetheless, 5 loci is measurably greater than the 2 which are minimally necessary to encode HSF and its activating factor. Some of these loci could encode additional factors that might act in a regulatory cascade culminating in induction of hsp transcription. Alternatively, they could act to interfere with the normal regulatory system through alterations in basic cellular physiology. It will be challenging to distinguish between the two possibilities.

REFERENCES

(1) Wu, C. (1984). Two protein-binding sites in chromatin implicated in the activation of heat shock genes. Nature 309: 229-234.
(2) Parker, C. S. and J. Topol (1984). A Drosophila RNA polymerase II transcription factor binds to the regulatory site of an hsp70 gene. Cell 37: 273-283.
(3) Zimarino, V. and C. Wu (1987). Induction of sequence-specific binding of Drosophila heat shock activator protein without protein synthesis. Nature 327: 727-730.
(4) Sorger, P. K., M. J. Lewis, and H. R. B. Pelham (1987). Heat shock factor is regulated differently in yeast and HeLa cells. nature 329: 81-84.
(5) Bonner, J. J., C. Parks, J. Parker-Thornburg, M. A. Mortin, and H. R. B. Pelham (1984). The use of promoter fusions in Drosophila genetics: Isolation of mutations affecting the heat shock response. Cell 37: 979-991.
(6) Parker-Thornburg, J. and J. J. Bonner (1987). Mutations that induce the heat shock response of Drosophila. Cell 51: 763-772.
(7) Hiromi, Y., and Y. Hotta (1985). Actin gene mutations in Drosophila: heat shock activation in the indirect flight muscles. EMBO J. 4: 1681-1687.
(8) Okamoto, H., Y. Hiromi, E. Ishikawa, T. Yameda, K. Isoda, H. Maekawa, and Y. Hotta (1986). Molecular characterization of mutant actin genes which induce heat shock proteins in Drosophila flight muscles. EMBO J. 5: 589-596.
(9) see discussions by Wiederrecht and by Pelham in this volume.
(10) Dura, J.-M. (1981). Stage dependent synthesis of heat shock induced proteins in early embryos of Drosophila melanogaster. Mol. Gen. Genet. 184: 381-385.

III. CELLULAR RESPONSES

Stress-Induced Proteins, pages 107–116

HEAT SHOCK IN *TETRAHYMENA* INDUCES THE ACCUMULATION OF A SMALL RNA HOMOLOGOUS TO EUKARYOTIC 7SL RNA AND *E. COLI* 4.5S RNA

Richard L. Hallberg and Elizabeth M. Hallberg

Department of Zoology, Iowa State University,
Ames, Iowa 50011

ABSTRACT We have examined the metabolism of a small stress-induced RNA in the ciliate *Tetrahymena thermophila* during both heat shock and starvation. During its rapid accumulation following administration of either stress, this RNA (G8 RNA) became associated with the large ribosomal subunit. This occurred at a time when changes in the translational properties of the cell are known to occur. This 306 nucleotide-long RNA was found to have sequence similarity to the eukaryotic 7SL RNA of the signal recognition particle and to the *E. coli* 4.5S RNA, both of which are known to be intimately associated with the translational machinery.

INTRODUCTION

A heat shock administered to the ciliate *Tetrahymena thermophila* induces the rapid accumulation of a small molecular weight RNA (G8 RNA) that becomes ribosome and/or polysome associated (1). This association takes place at a time when the translational regulatory properties of the cell are known to be changing (1,2). Additionally, in contrast to other genes whose transcriptions are known to be induced or elevated by heat shock, the gene for G8 RNA is apparently transcribed by RNA polymerase III rather than RNA polymerase II (1). Because of the distinct possibility that G8 RNA may be a structural RNA that plays some role in the regulation of protein synthesis which occurs in heat shocked cells, we have further characterized the control of its synthesis

This work was supported by NSF grant PCM-850794

and accumulation in both heat shocked and otherwise stressed cells. In addition, its stucture and ribosome association have been analyzed in more detail. Finally, in comparing the sequence of this RNA with other small, stable RNAs, we find that the nucleotide sequence of G8 RNA is similar to two other RNAs, viz., *E. coli* 4.5S RNA (3) and eukaryotic 7SL RNA (4), both of which are known to associate with ribosomes and to be involved in some way in the modification or regulation of protein synthesis.

RESULTS

Synthesis and Accumulation of G8 RNA

The accumulation of G8 RNA is induced not only by heat shock but by starvation as well (1). To assess both the quantitative and qualitative similarities of these inductions, we measured the relative accumulation kinetics of G8 RNA in cells exposed to these two stresses and found (fig. 1) that in both cases it accumulated rapidly to a maximum level per cell and then remained essentially constant (heat shocked cells) or gradually declined over time (starved cells). In the latter case, this decline followed the general decline in overall RNA levels observed in starved cells (5). Thus, in terms of percentage of total RNA, the maximum level of G8 RNA in starved cells also remained essentially constant. In both cases, the amounts of G8 RNA in cells which were returned to normal growth conditions after having been starved or heat shocked declined to <1% of their peak levels in less than 30 min (data not shown).

We had previously estimated (1), using indirect methods, that the absolute amount of G8 RNA in heat shocked cells was such that there might be one G8 RNA per polysomal ribosome in cells heat shocked for 90 min. Were it correct, such a finding could have important functional implications. Consequently, we made more direct and accurate measurements of the maximum G8 RNA levels in heat shocked cells as well as in starved cells. Two methods were used. In the first, we subcloned our G8 cDNA (1) into an expression vector (pBS; Vector Cloning Systems) from which we could prepare "sense" and "antisense" transcripts of the G8 RNA. We prepared radioactively labelled copies of "antisense" G8 RNA and hybridized them, in excess, and to completion, to varying amounts of RNA extracted from cells heat shocked at 40°C for 30 min, from cells starved for 1.5 hr, or from cells in early log phase of growth at 30°C (table 1). This procedure of single-stranded probe-excess titration is the most accurate way of determining the absolute cellular abundances of nucleic acid species (6). As a complementary and independent method for determining the abundance of G8 RNA, we prepared non-radioactive "sense" G8 RNA using our expression vector and added this RNA in varying, known amounts to total RNA samples isolated from log phase cells, which contain no detectable G8 RNA. Using a 25 base oligonucleotide complementary to an internal stretch of the G8 RNA molecule, we then performed primer extensions on these RNA samples and on RNA isolated from 1 hr heat shocked cells. By assessing

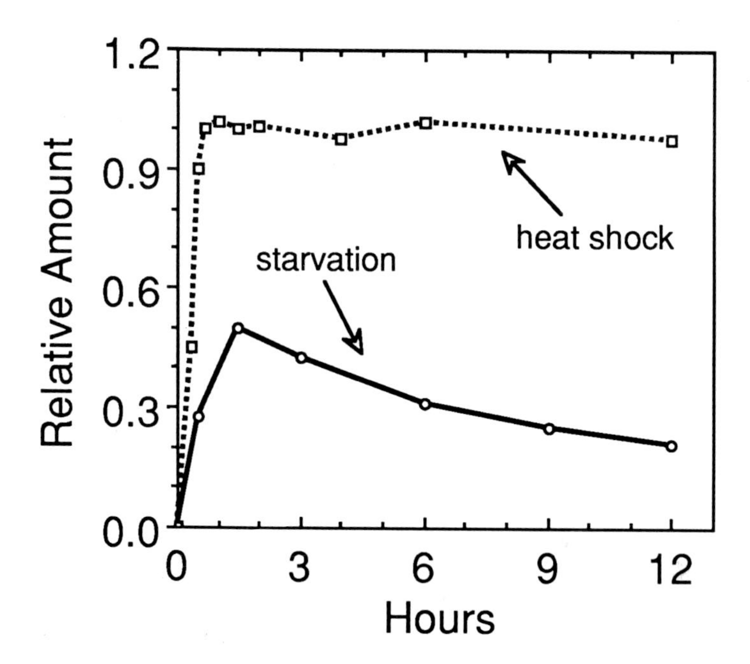

FIGURE 1. Accumulation of G8 RNA during heat shock and starvation. Cells in log phase growth at 30°C were either tranferred to 40°C or washed into starvation medium. At various times after this, cells were isolated and their RNA was extracted. Dilution series of cell equivalents of each RNA sample were applied to a membrane filter using a slot blot apparatus. All filters were then simultaneously hybridized with an in vitro labelled anti-G8 transcript. Quantitation of the hybridized RNA was done by densitometric analysis of the resultant autoradiograms.

that amount of G8 "sense" RNA added to RNA from non-stressed cells which gave a primer extension signal equal to that obtained with RNA from heat shocked cells, we estimated that the percentage of G8 RNA in heat shocked cells was between 0.025 and 0.05% of the total cellular RNA, a value in close agreement with our hybridization measurement of 0.041% (table 1).

TABLE 1
ABSOLUTE CELLULAR LEVELS OF G8 RNA

	% of total RNA[a]	number of molecules/cell[a]
heat shocked cells (30 min at 40 °C)	0.0410	85,000
early-mid log phase cells	<0.0007	<150
1.5 hr starved cells	0.0210	43,500

[a] In all cases radioactive transcripts of antisense G8 RNA of known specific activity were hybridized in solution with varying, known quantities of each of the RNA samples. After S1 digestion of the completed hybridization reactions, the TCA precipitable counts remaining were used to calculate the absolute number of molecules using the method described by Davidson (6).

These findings indicated that our earlier estimate (1) of G8 abundance in heat shocked cells was off by about a factor of ten. Clearly, these cells can not possess a G8 RNA molecule on each polysomal ribosome. At best, there can be one G8 RNA per polysomal message. Whether that is in fact the actual distribution of these RNAs remains to be determined.

Another finding of the primer extension experiment described above was that we could accurately determine the 5' end of the in vivo G8 RNA. From this we concluded that the normal in vivo G8 RNA was 306 nucleotides in length.

We have determined that there is a single copy of the G8 gene per haploid genome (data not shown). As the micronucleus of *T. thermophila* is 45 ploid (7), the minimum transcriptional output of these genes necessary to produce the approximately 85,000 copies of G8 RNA in the first 30 min of heat shock is about 63 transcripts per gene per minute. A similar calculation (not shown) for the minimum transcription rate per gene necessary to produce 5S rRNA in cells growing at their maximum growth rate is 65 transcripts per gene per minute. Thus, a single genomic copy of the RNA polymerase III - transcribed G8 gene, if transcribed at a maximum rate, can easily account for the accumulation rate observed.

Ribosome Association of G8 RNA

We had shown (1) that at 30 min of heat shock G8 RNA was found primarily (>90%) associated with monosomic ribosomes and that later, at 90 min of heat shock, most of the G8 RNA was found associated with polysomes. A similar type of distribution of G8 RNA association was found in starved cells (data not shown), with the difference being that a lesser percentage of the total G8 RNA ultimately becomes polysome associated. The majority of the G8 RNA in the cell is always found in the mono- and di-somal ribosome peaks. To determine with what component or components of the translational machinery G8 RNA was associating, a mixture of all polysomal and non-polysomal ribosomes were prepared from 90 min heat shocked cells, dissociated into subunits by salt, and then run on sucrose gradients (fig. 2a). Under these conditions about 70% of the G8 RNA was found to cosediment with the large ribosomal subunit while the other 30% sedimented in a macromolecular form sedimenting at about 20-30S. When an identical ribosome preparation was dissociated into subunits by lowering the Mg^{++} concentration, all the G8 RNA was found to be sedimenting in the 20-30S region of the gradient (fig. 2b). What else might be in this macromolecular complex remains to be determined, and even whether this complex is an artifact of the isolation procedure must also be ascertained. Nonetheless, when we prepared ribosomal subunits from cells isolated early or late in starvation or early or late in heat shock using either of the above procedures, we found identical patterns of G8 RNA association and sedimentation. Regardless of whatever else G8 RNA may be associated with, and whether it is predominantly monosomally or polysomally distributed, it appears from these analyses that G8 RNA, and whatever else it may be complexed with, always has a primary interaction with the large ribosomal subunit.

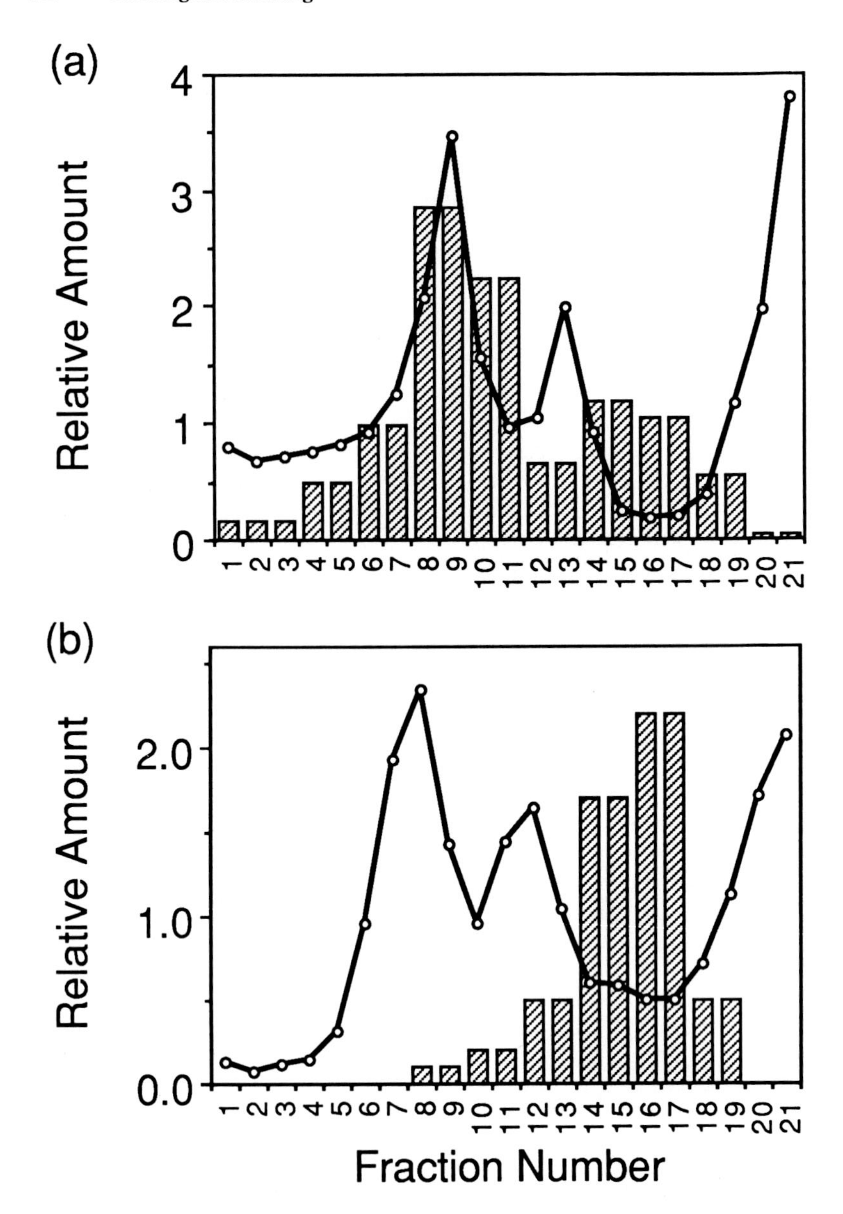

FIGURE 2. Effect of ribosome dissociation on the sedimentation characteristics of G8 RNA. The total ribosome population was prepared from cells heat shocked for 90 min at 40°C. These ribosomes were then dissociated into subunits by

Sequence Homologies With Other Small RNAs.

One other RNA polymerase III transcript found in eukaryotic cells which is approximately 300 nucleotides long is 7SL RNA, a structural and functional component of the signal recognition particle, a ribonucleoprotein complex (4, 8). This particle is known to interact with ribosomes via the large subunit and, furthermore, it possesses translational regulatory properties (4, 8). Because of these similarities to what we had found (and surmised) regarding G8 RNA, we compared the sequences of *T. thermophila* G8 RNA with *Xenopus laevis* 7SL RNA and found that a reasonably long stretch of homology was apparent within the middle of the two RNAs (fig 3a). Outside this region of reasonably high sequence similarity the homology between the two RNAs is no better than random (fig. 3b). Interestingly, and of possible functional significance, nucleotide stretches of potential intramolecular base pairing in the G8 RNA sequence (data not shown) are very similar to what has been described for 7SL RNA (9).

Another small molecular weight RNA which has recently been shown to interact with ribosomes via elongation factor G (which therefore interacts with the large ribosomal subunit) and that is required for protein synthesis, is the *E. coli* 4.5S RNA (10). A sequence comparison of G8 RNA with 4.5S RNA indicated (fig. 4) that with the exception of the approximately first and last 10 bases, the remainder of the 4.5S RNA displayed a 54% overall sequence similarity to G8 RNA, and there were stretches of 41 and 20 bases showing > 65% correspondence. Of note, the regions of greatest similarity between G8 and 4.5S were also regions of high sequence homology between G8 and 7SL RNA (fig. 3a).

---

either dissolving them in a buffered solution containing 0.5 M KCl and 0.005 M $MgCl_2$ (a), or in a buffered solution containing 0.1 M KCl and no Mg^{++} (b), and then centrifuging them on sucrose gradients made up in the same solutions. Fractions were collected and their absorbances determined (o). Adjacent 2 or 3 fractions were pooled, their RNAs extracted, electrophoresed on agarose gels, transferred to membrane filters, and then hybridyzed to a G8 RNA probe. The relative degree of hybridization to these fractions is indicated by the height of the bars. The bottom of the gradient is on the left.

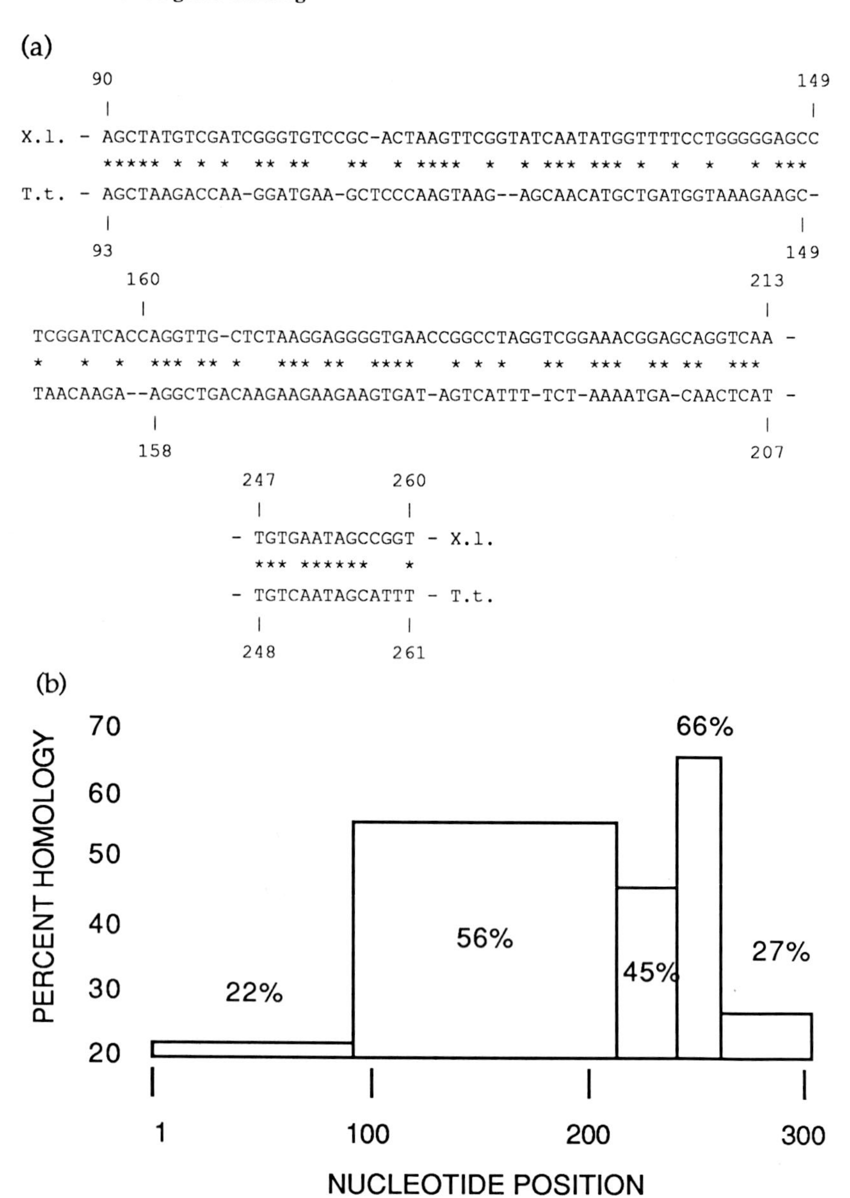

FIGURE 3. Sequence comparisons of *Tetrahymena thermophila* (T.t) G8 RNA and *Xenopus laevis* (X.l.) 7SL RNA genes. (a) regions showing greatest homology. (b) distribution of sequence homology.

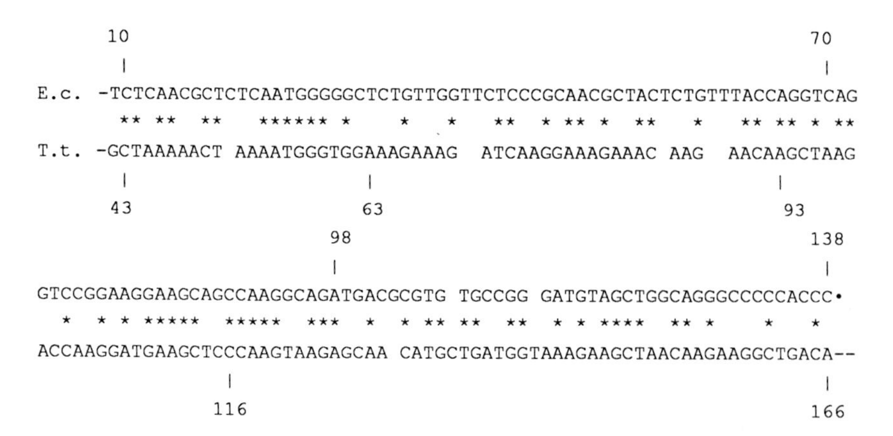

FIGURE 4. Sequence comparison between *T. thermophila* (T.t) G8 RNA and *E. coli* (E.c.) 4.5S RNA genes.

Discussion

Heat shock elicits two changes in the protein synthetic properties of *T. thermophila* cells: the ability to selectively translate mRNAs changes (1), and protein synthesis becomes thermostabilized (2). Although the accumulation of G8 RNA coincides with that time in heat shock when the above changes are occurring, and that, in addition, it interacts with ribosomes indicate the possibility that G8 RNA might participate in some aspect of the regulation of protein synthesis, there is no direct evidence that this is so. One way we have been approaching a more definitive resolution of this problem is to identify mutant strains which exhibit abnormal protein synthetic properties during heat shock. One strain we have identified and characterized is unable to thermostabilize protein synthesis but is still able to alter its translational discriminatory properties (2). The accumulation and ribosome association of G8 RNA in this strain is indistinguishable from that seen in wild type cells (data not shown). Thus, if G8 RNA does affect translation in some way, it appears that it must be with regard to the selectivity of mRNA utilization. A more rigorous investigation of this possibility is currently being pursued.

Should G8 RNA be involved in translational regulation, it might function either as a positive or negative regulator. Were it to function in a signal recognition-like particle (SRP) it could exert its effect by inhibiting further heat shock mRNA translation, thus permitting the return of non-stress mRNA utilization. In the same way that the SRP recognizes the nascent signal peptide and then inhibits further elongation, a G8 RNA containing particle may recognize the nascent heat shock proteins and prevent further synthesis. However, that region of the 7SL RNA known to be involved in elongation arrest is that portion which has no similarity to G8 RNA. With regard to the possibility of a positive regulator, as 4.5S RNA is a necessary component of the translational apparatus, it could be that G8 RNA is required for the translation of non-stress mRNAs at high temperatures and that it replaces another similar but non-identical RNA which functions at non-stress temperatures. Both hypotheses make certain predictions which we are currently in the process of testing.

REFERENCES

1. Kraus KW, Good PJ, Hallberg RL (1987). Proc. Nat. Acad. Sci. USA 84: 383-387.
2. Kraus KW, Hallberg EL, Hallberg RL (1986). Mol. Cell. Biol. 6: 3854-3861.
3. Hsu LM, Zagorski J, Fournier MJ (1984). J. Mol. Biol. 178: 509-531.
4. Walter P, Blobel G (1982). Nature 299: 691-698.
5. Sutton CA, Hallberg RL (1979). J. Cell. Physiol. 101: 349-358.
6. Davidson EH (1986). "Gene Activity in Early Development." Orlando: Academic Press, pp. 541-543.
7. Woodard J, Kaneshiro ES, Gorovsky MA (1972). Genetics 70: 251-260.
8. Walter P, Gilmore R, Blobel G (1984). Cell 38: 5-8.
9. Zweib C (1985). Nucleic Acids Res. 13: 6105-6124.
10. Brown S (1987). Cell 49: 825-833.

Stress-Induced Proteins, pages 117–128

ANION TRANSPORT IS LINKED TO HEAT SHOCK INDUCTION[1]

Ann C. Sherwood, Kathleen John-Alder and Marilyn M. Sanders

Department of Pharmacology, UMDNJ-Robert Wood Johnson Medical School, Piscataway, New Jersey 08854

ABSTRACT Certain drugs which block anion transport systems in the cell membrane induce the changes in gene expression and cell growth typical of the heat shock response in *Drosophila* Kc cells. Characterization of anion uptake in these cells shows there are two major transport systems, one of which shows reduced activity at temperatures where heat shock is induced. Somatic cell variants selected for their resistance to growth inhibition by anion transport blockers of the disulfonic stilbene class show decreased heat shock polypeptide induction, changes in growth properties, changes in anion transport kinetics and amplified polypeptides, one of which binds an anion transport blocker.

INTRODUCTION

In studies designed to assess the effects of medium composition on the heat shock response in *Drosophila* Kc cells, we observed that heat shocked cells are hypersensitive to changes in solutes and osmolarity of the growth medium. This suggested that the heat shock response could be perturbed by inhibitors of ion transport systems in the cell membrane which participate in cell volume or growth regulation in eukaryotic cells (1). Tests of a number of classes of ion transport inhibitors on normal and heat shocked *Drosophila* cells showed that 4,4'diisothiocyano-2,2'stilbene disulfonate (DIDS), an impermeant anion transport blocker, induced heat shock polypeptide synthesis at normal temperatures. Inhibitors of other transport systems did not show this effect. The effects of DIDS on anion transport have

[1]This work was supported by NIH GM 34741.

been characterized in many cell types (2) and the protein mediating DIDS-sensitive anion exchange in the red blood cell has been studied in some detail (3). We have begun to investigate the effects of DIDS on anion transport in Kc cells in hopes that this may lead to the identification of physiological mechanisms which are important to the induction and regulation of the heat shock response.

RESULTS

Figure 1 shows the pattern of ^{35}S-methionine incorporation into protein in cells at 25° C (normal), cells at 37° C (heat shock) and cells which have been treated with 10^{-4} M DIDS at 25° C. The DIDS-treated cells show significant induction of all the *Drosophila* heat shock polypeptides (hsps) and this is accompanied by the shutdown of normal protein synthesis typically seen in this system. The time course of DIDS induction of hsp synthesis is slower than heat induction (15-30 min compared with <5 min, respectively), but compared with the response induced by other nonspecific drugs in *Drosophila* (*e.g.*, canavanine or arsenite) the DIDS-induced effect is faster and more complete (4).

To further characterize the nature of the DIDS-induced heat shock response we have tested the effects of DIDS on a number of physiological changes associated with heat shock. These studies have shown first, as illustrated by Figure 1, that the transcriptional and translational changes typical of heat shock are caused by DIDS. DIDS also inhibits incorporation of ^{3}H-thymidine into DNA (5), causes a collapse of the intermediate filament cytoskeleton onto the side of the nucleus (6, data not shown), and partially dephosphorylates the ribosomal protein S6 (4, data not shown). The heat shock-induced changes in mRNA processing and thermotolerance (7) have not yet been tested. These comparisons indicate that this drug causes most, if not all, of the physiological changes seen in heat-stressed *Drosophila* cells.

Other anion transport inhibitors have also been tested to determine whether they cause hsp induction. This control is necessary because DIDS covalently reacts with its target protein and irreversibly inhibits transport. At high concentrations, it is also known to react nonspecifically with other macromolecules on the surface of the cell (8). Thus, it was important to determine whether anion transport inhibitors which do not have reactive substituents also cause a heat shock response. Figure 1D shows that flufenamic acid

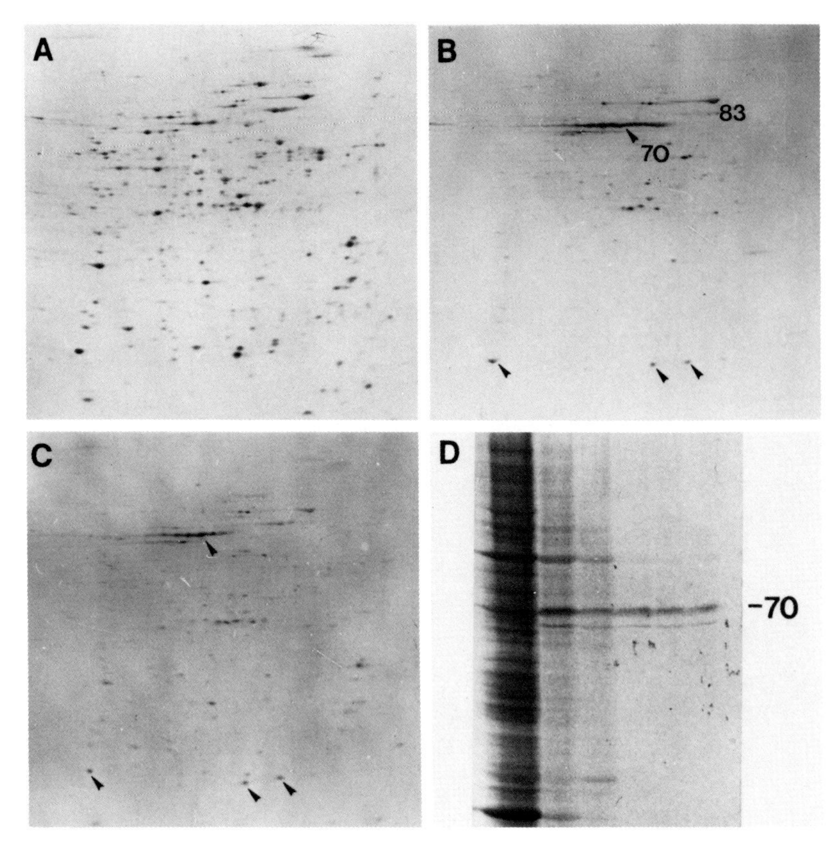

FIGURE 1. Autoradiograms of SDS acrylamide gels of polypeptides labeled by ^{35}S-methionine in (A) normal conditions at 25° C, (B) in heat shock at 37° C and (C) in 10^{-4} M DIDS at 25° C. (D) shows labeling of cells treated with 1) no FFA, 2) 50 μM, 3) 75 μM, 4) 100 μM, 5) 125 μM and 6) 150 μM FFA at 25° C. The cells were treated for 60 min before labeling for an additional 30 min. Labeling and electrophoresis conditions have been described previously (10).

(FFA), an anthranilic acid derivative which inhibits anion transport in the red blood cell (9) and in Kc cells (see below), induces heat shock polypeptides.

To attempt to determine whether the anion transport inhibitors, DIDS and FFA, were acting through physiological functions related to anion transport, anion transport was characterized in Kc cells. This showed that uptake of $^{36}Cl^-$ is rapid in these cells, approaching a steady state in two min (11). The steady state chloride content measured both with $^{36}Cl^-$ and with a Cl^- specific electrode was found to vary with the medium chloride concentration ($[Cl^-]_o$) between 15 nmoles/10^7 cells at 4 mM Cl^-_o and 95 nmoles/10^7 cells at 68 mM Cl^-_o (11). Since the cells synthesize protein and mount a normal heat shock response with widely varying Cl^- contents, it is likely that total cellular Cl^- content *per se* has little to do with the heat shock response. This does not rule out some other role for an anion transport system in heat shock, however.

Measurement of $^{36}Cl^-$ uptake with variation of $[Cl^-]_o$ showed kinetic evidence for the presence of a minimum of two uptake systems, one with a relatively high affinity for Cl^- and reaching a steady state at about 15 mM Cl^-_o. The second has a lower affinity for Cl^- and appears at Cl^-_o above 30 mM (11). A summary of the relative sensitivities of these two uptake systems to a series of inhibitors and other anions is shown in Table 1.

TABLE 1
EFFECTS OF VARIOUS AGENTS ON CHLORIDE UPTAKE SYSTEMS

Effector	High Affinity Uptake[a]	Uptake at 68 mM Cl^-[a]
10 μM DIDS	0.5	0.95
50 μM DIDS	0.35	0.85
50 μM FFA	0.3	0.74
6 mM HCO_3^-	0.3	>0.9
20 mM HCO_3^-	0.05	0.9
35 μM VO_4^{3-}	0.95	0.65
12 mM NO_3^-	0.7	0.4

[a]Fraction normal uptake activity. High affinity uptake measured at $[Cl^-]_o$ < 20 mM. Uptake at 68 mM Cl^- reflects sum of effect on activity of both high and low affinity systems. Data from (11).

Using these characterizations to distinguish between the two chloride uptake systems, we have measured $^{36}Cl^-$ uptake as a function of temperature. Figure 2 shows that the high affinity system becomes less active at temperatures higher than 25°C. In other uptake studies, we have found that

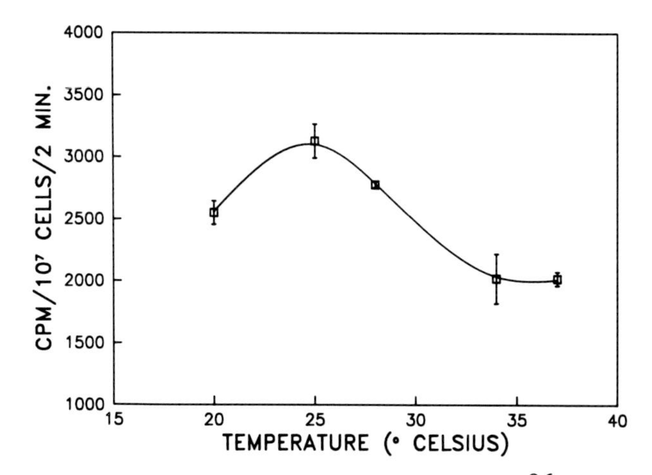

FIGURE 2. Temperature-sensitivity of $^{36}Cl^-$ uptake through the high-affinity uptake system. The cells were incubated for 60 min in a 4 mM Cl^- balanced salt solution (BSS), conditions where only the high affinity uptake system is active (11), at the temperatures indicated. $^{36}Cl^-$ uptake was measured as described elsewhere (11).

canavanine and arsenite, drugs which induce hsp synthesis in Kc cells, also cause a similar decrease in $^{36}Cl^-$ uptake via the high affinity uptake system (data not shown). The finding that heat as well as drug-induced heat shock responses all result in decreased anion uptake circumstantially supports the possibility that the activity of the high affinity anion uptake system plays a physiological role in the heat shock response.

An independent approach to studying the role of transport systems in heat shocked cells has involved selecting somatic cell variants altered in their ability to take up anions. As discussed above, DIDS and related anion transport inhibitors inhibit DNA synthesis (5) and cell growth is arrested in the presence of the drugs. We used a step-up selection protocol (12) to obtain Kc cell variants resistant to the growth inhibiting properties of DIDS and have characterized the phenotypes of the variants with respect to ability to take up anions, the amount of induction of hsps and the protein amplification patterns.

Figure 3 shows that the variants have increased capacity for the uptake of $^{36}Cl^-$. Both the variant cells maintained

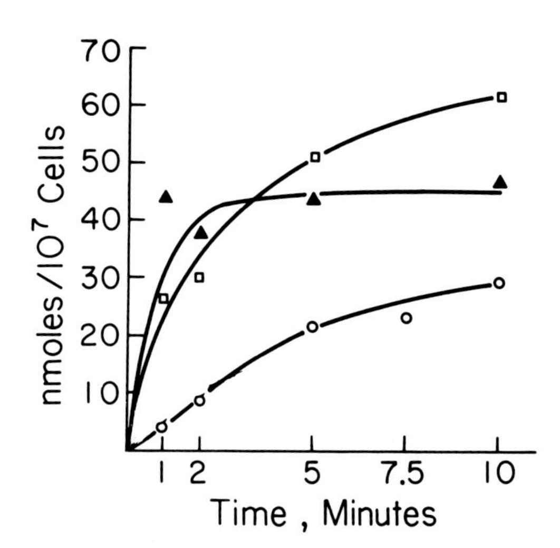

FIGURE 3. Time course of anion uptake in the DIDS-resistant variants. $^{36}Cl^-$ uptake was measured as a function of time as described elsewhere (11). (O) EC-1 parent; (□) P-20 variant growing in the presence of DIDS; (▲) RB-1 variant growing in the absence of DIDS.

in the presence of the DIDS, designated P-20, and cells which have been removed from selective conditions, designated RB-1, take up $^{36}Cl^-$ significantly more rapidly than the parent clone, EC-1. This shows that the variants have more available anion uptake systems than the parent, as expected from the fact they were selected for resistance to an anion uptake inhibitor.

Because DIDS causes the induction of hsp synthesis, the characteristics of the heat shock response in the variants were of interest. The levels of expression of the *Drosophila* hsps in the parent line, EC-1, and in two lines of variant cells which had been removed from DIDS selection at different times were determined. When tested in this experiment, RB-1 had been out of DIDS selective conditions for 30 weeks and had lost most of the growth resistance phenotype, the increased anion uptake phenotype and the protein amplification phenotype present in the variant, P-20. The second variant line, RB-2, had been out of DIDS selective conditions for 4 weeks and had anion uptake properties similar to those seen for RB-1 in Figure 3. Growth resistance to DIDS and protein amplification patterns were still at maximal levels. Figure

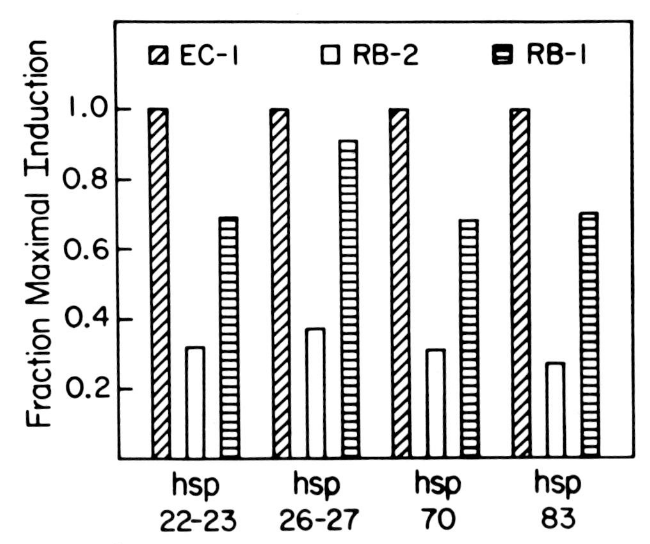

FIGURE 4. Expression of hsps in the DIDS-resistant variants. Equal numbers of EC-1 (parent), RB-2 (variant 4 weeks out of DIDS selection) and RB-1 (variant 30 weeks out of DIDS selection) were labeled with ^{35}S-methionine for 60 min at 37° C. Autoradiograms of SDS acrylamide gels of whole cell proteins (10) were scanned in the linear range of the film with a Joyce Loebls integrating densitometer. The level of expression for each hsp in the parent cell line is defined as 1.0.

4 shows that RB-2 shows only 30% of the hsp expression seen in the parent EC-1 while RB-1 has reverted to a level of hsp expression close to that seen for EC-1. Thus the DIDS-resistant variants have lowered levels of expression of hsps.

The DIDS resistant variants show dramatic amplification (8 to >16-fold) of expression of three polypeptides which can be readily detected on stained polyacrylamide gels (Sherwood *et al*., ms submitted). The amplified polypeptides are associated with the detergent insoluble cytoskeleton and are not expressed in heat shock conditions. They have molecular weights of 46, 62 and 116 kD.

Since the variants were selected for resistance to a drug which blocks an anion transport system in the cell membrane we expected that an exterior membrane protein should be amplified to account for the increased anion uptake seen (Fig. 3). To determine whether the variants had amplified putative anion transport proteins, the cells were labeled on the outside with sulfo-succinimidyl biotin (sulfo-NHS-biotin) (13). The reacted exterior membrane proteins were detected on nitrocellulose blots of discontinuous SDS acrylamide gels by reaction with ^{125}I-avidin (13). Figure 5 shows, in the leftmost two lanes, the polypeptides labeled in gels of solubilized whole parent (lanes E) and variant (lanes R) cells. The variants show substantial amplification of one exterior membrane polypeptide of molecular weight 116 kD. The pair of lanes labeled 'cyt' show that this 116 kD membrane polypeptide copurifies with detergent insoluble cytoskeletons purified from the sulfo-NHS-biotin reacted cells.

To determine whether this 116 kD amplified membrane protein could be an anion transporter, a drug protection experiment modified from the above-described labeling experiment was performed. The cells were treated with the reversible anion transport inhibitor, FFA, and were labeled in the presence of FFA with sulfo-NHS-acetate, a reagent which reacts with all the unprotected reactive groups on the outside of the cell. Following the first reaction period, the FFA and sulfo-NHS-acetate were washed out and the cells were treated with sulfo-NHS-biotin to label sites protected by the FFA. Following treatment of the nitrocellulose blots with ^{125}I-avidin the patterns on the right lanes of Figure 5 were seen. The FFA protected the same 116 kD band shown to be amplified in the whole membrane labeling experiment. In the protection experiment this protein can be easily seen in the parent Kc (lanes K) and EC-1 (lanes E) samples and it is greatly amplified in the variant (lanes R) cells. As with the whole membrane labeling results, this 116 kD polypeptide copurifies with the detergent insoluble cytoskeleton.

DISCUSSION

We have shown that DIDS (a disulfonic stilbene) and FFA (an anthranilic acid derivative) induce heat shock polypeptide synthesis in _Drosophila_ Kc cells (Fig. 1). These agents also inhibit $^{36}Cl^-$ uptake in these cells (Table I). Partial characterization of the $^{36}Cl^-$ uptake systems in Kc cells showed that two major systems are present. These differ, first, in activity in response to $[Cl^-]_o$ and thus have been designated as "high" and "low" affinity systems.

The high affinity system shows steady state uptake at $[Cl^-]_o$ >15 mM and is substantially inhibited by DIDS and FFA as well as by increasing concentrations of HCO_3^- and NO_3^-.

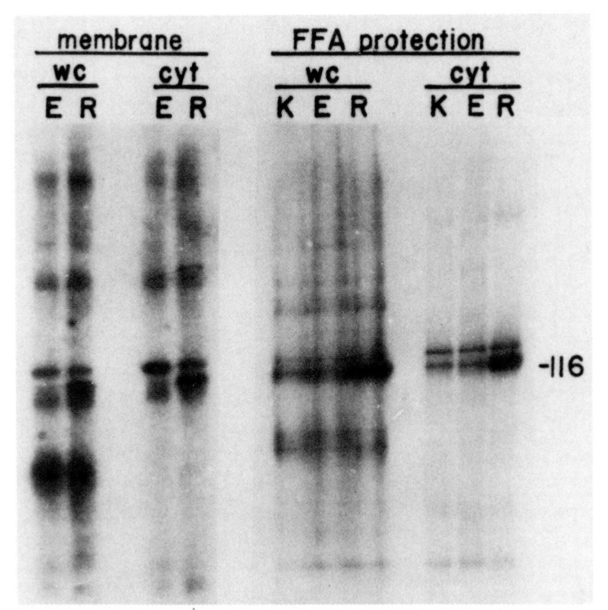

FIGURE 5. Amplified membrane and FFA binding proteins in the variants. Membrane--intact cells were labeled as described in the text with sulfo-NHS-biotin (13). FFA protection--intact cells were labeled first with sulfo-NHS-acetate in the presence of 100 μM FFA and then with sulfo-NHS-biotin as described in the text. WC, autoradiograms of gels of whole cells; CYT, of cytoskeletons; lanes E, EC-1 parent; R, RB-2 variant; K, Kc cell line. The 116 kD polypeptide is marked.

The effects of HCO_3^- suggest the high affinity system may be a Cl^-/HCO_3^- exchanger. Experiments designed to determine whether this transport system is involved in intracellular pH regulation are in progress. The transport activity of the high affinity system can be studied in isolation by measuring uptake at $[Cl^-]_o$ below 20 mM.

Measurement of $^{36}Cl^-$ uptake through this system as a function of temperature shows uptake activity is maximal at the normal growth temperature of 25° C and decreases at temperatures of 28° C and above where hsp synthesis is induced (Fig. 2) (7). Anion transport through the high affinity system is also reduced in canavanine and arsenite treated Kc cells (data not shown). Thus, reduced chloride uptake through the high affinity chloride uptake system is consistently observed in the presently tested cases of heat and drug induced hsp production.

The low affinity chloride uptake system is activated at $[Cl^-]_o > 30$ mM and is not significantly inhibited by DIDS or FFA (11). Neither is its activity affected by HCO_3^- but it is inhibited by VO_4^{3-} and in a concentration independent manner by NO_3^- (11). The activity of the low affinity uptake system simply increases as temperature is increased in the normal growth and heat shock induction range (data not shown).

The relationship between anion transport inhibition and hsp induction has also been investigated in somatic cell amplification variants, selected for resistance to the growth inhibiting properties of DIDS (Sherwood et al., ms submitted). These cells have increased capacity for $^{36}Cl^-$ uptake (Fig. 3), suggesting they have amplified expression of a plasma membrane anion transport system. The exterior cell labeling experiment in Figure 5 shows this is indeed the case. Furthermore, a protection experiment shows the 116 kD amplified membrane protein binds FFA and identifies it as a putative anion transporter. This protection protocol has not been used previously to identify anion transporters, but these experiments clearly show the technique is sufficiently sensitive for labeling even in normal cells where expression of this protein is not amplified.

The level of induction of hsp's in the DIDS-resistant cells is reduced to one-third the level seen in the parent cells (Fig. 4). Thus the pattern of response of hsp induction to the activity of the DIDS-sensitive anion uptake system is consistent with more anion transport capacity leading to less hsp production and lowered capacity leading to increased hsp production. Although the correlation is purely circumstantial at this time, it is consistent with the possibility that the DIDS sensitive anion uptake system in the cell membrane participates in regulation of the heat shock response.

The present correlations between anion transport function and regulation of hsp production provide two clues suggesting possible signalling systems which could play a role in communicating between the plasma membrane and sites of regulation of the heat shock response. The first, a role for intracellular pH, is suggested by the possibility that the DIDS-sensitive chloride uptake system is a Cl^-/HCO_3^- exchanger which participates in intracellular pH regulation in Kc cells. Intracellular pH has been shown to drop in heat shocked *Drosophila* salivary glands (14) and in yeast (15) but a regulatory role for these changes has not been demonstrated. A second possible signalling system suggested by our present data involves the detergent-insoluble cytoskeleton. A total of three insoluble cytoskeleton proteins are amplified in the DIDS-resistant variants. One is the 116 kD FFA-binding exterior membrane protein and the other two are components of the structure which forms a peri-nuclear cap in heat shocked cells (5,6,16). The possibility that the detergent-insoluble cytoskeleton plays a role in heat shock regulation is attractive because the time course of perinuclear cap formation follows the time course of induction of and recovery from the heat shock response more closely than any other physiological change known at the present time (16).

REFERENCES

1. Kregenow FM (1981). Osmoregulatory salt transporting mechanisms: Control of cell volume in anisotonic media. Ann Rev Physiol 43:493.
2. Hoffmann EK (1986). Anion transport systems in the plasma membrane of vertebrate cells. Biochim Biophys Acta 864:1.
3. Rothstein A (1984). The functional architecture of band 3, the anion transport protein of the red cell membrane. Can J Biochem and Cell Biol 62:1198.
4. Olsen AS, Triemer DF, Sanders MM (1983). Dephosphorylation of S6 and expression of the heat shock response in Drosophila melanogaster. Molec Cell Biol 3:2017.
5. Sanders MM, Feeney-Triemer D, Olsen AS, Farrell-Towt J (1982). Changes in protein phosphorylation and histone H2b disposition in heatshock in Drosophila. In Schlesinger MS, Ashburner M, Tissieres, A (eds): "Heat Shock from Bacteria to Man," Cold Spring Harbor: Cold Spring Harbor Laboratory, p 235.
6. Falkner FG, Saumweber H, Biessmann H (1981). Two Drosophila melanogaster proteins related to intermediate filament proteins of vertebrate cells. J Cell Biol 91:175.
7. Lindquist S (1986). The heat-shock response. Ann Rev Biochem 55:1151.
8. Cabantchik ZI, Rothstein A (1974). Membrane proteins related to anion permeability of human red blood cells. J Membr Biol 15:207.
9. Falke JJ, Chan SI (1986). Molecular mechanisms of band 3 inhibitors I. Transport site inhibitors. Biochem 25: 7888.

10. Sanders MM (1981). Identification of histone H2b as a heat shock protein in Drosophila. J Cell Biol 91:579.
11. Sherwood AC, John-Alder, Sanders MM (1988). Characterization of chloride uptake in Drosophila Kc cells. J Cell Physiol (in the press).
12. Schimke RT (1984). Gene amplification in cultured animal cells. Cell 37:705.
13. Ingalls HM, Goodloe-Holland CM, Luna EJ (1986). Junctional plasma membrane domains isolated from aggregating Dictyostelium discoideum amebae. Proc Nat Acad Sci USA 83:4779.
14. Drummond IAS, McClure SA, Poenie M, Tsien RY, Steinhardt RA (1986). Large changes in intracellular pH and calcium observed during heat shock are not responsible for the induction of heat shock proteins in Drosophila melanogaster. Molec Cell Biol 6:1767.
15. Weitzel G, Pilatus U Rensing L (1987). The cytoplasmic pH, ATP content and total protein synthesis rate during heat-shock protein inducing treatments in yeast. Exp Cell Res 170:64.
16. Sanders MM, Triemer DF, Brief B (1985). Heat shock causes a rearrangement in the intermediate filament cytoskeleton in Drosophila. Ann NY Acad Sci 455:710.

Stress-Induced Proteins, pages 129–135

A HEAT SHOCK PROMOTER-DRIVEN ANTISENSE RIBOSOMAL PROTEIN GENE DISRUPTS OOGENESIS IN DROSOPHILA[1]

Su Qian, Seiji Hongo, and Marcelo Jacobs-Lorena

Case Western Reserve University, Department of Developmental Genetics and Anatomy, Cleveland, Ohio 44106

ABSTRACT. We constructed an antisense rpA1 ribosomal protein gene driven by the strong and conditional heat shock hsp70 promoter. This construct was inserted into the germ line of Drosophila via P-element mediated transformation. Induction of the antisense gene in tranformed flies had a strong effect on oogenesis, causing the formation of abnormal and unfertile eggs. Thus, females bearing the antisense rpA1 gene behaved as conditional female sterile mutants. This phenotype is likely to be caused by a general destabilization of cellular mRNAs upon induction of the antisense gene and to the failure of nurse cells to transfer their contents to the oocyte at the end of oogenesis.

INTRODUCTION

The demand for new ribosomes during Drosophila oogenesis is extremely high. Although most of the 5 X 10^{10} ribosomes stored in the Drosophila egg are synthesized in a matter of hours, it takes weeks to synthesize the 9 X 10^{7} ribosomes of the mouse egg and months to synthesize the 7 X 10^{8} ribosomes of the sea urchin egg. In Drosophila, a large output of ribosomes may be required at other developmental periods as well, as suggested by the phenotype

[1]This research was supported by a grant from the National Institutes of Health.

of mutants that affect ribosome synthesis (e.g. bobbed, Minute, mini). Formation of the adult bristles and abdominal coverings during pupation is defective in these mutants (reviewed in ref. 1).

Minutes are a class of approximately 50 different recessive lethal Drosophila genes whose dominant phenotype is similar to the phenotype of bobbed and mini. For this and other reasons, Minutes have been hypothesized to correspond to lesions in ribosomal protein genes. In only one case has a Minute been demonstrated to correspond to a ribosomal protein (rp49) gene (2).

We have recently isolated and characterized the gene coding for the Drosophila ribosomal protein rpA1 (3). rpA1 is a single copy gene and is homologous to the "A" type eukaryotic ribosomal proteins that appear to play an essential function in initiation and elongation of protein synthesis. rpA1 maps to a poorly characterized region of the second chromosome and our attempts to correlate it with the M(2)S7 Minute mutation, that maps to the same area, were unsuccessful (Qian and Jacobs-Lorena, unpublished results). To assess the effect of the reduction of rpA1 gene expression has on development, we constructed and transformed into the fly's germ line an antisense copy of the rpA1 gene driven by the strong and conditional heat shock hsp70 promoter. We found that expression of the antisense rpA1 gene severely disrupts oogenesis but has little or no effect on other developmental processes.

RESULTS

Induction of rpA1 Antisense Gene Expression Severely Disrupts Oogenesis

Constructs that placed the rpA1 gene either in sense or antisense orientation under control of the hsp70 promoter were transformed into the germ line of Drosophila via P-element mediated transformation. Three independent antisense fly lines and one control sense line (hereafter termed "antisense flies" and "sense flies", respectively) were studied in detail. Surprisingly, even very frequent induction of the antisense gene (1-hour pulses at 37°C given every 6 hours) throughout postembryonic development had no noticeable effect. However, antisense gene expression severely disrupted oogenesis.

Antisense females (but not sense females) produced eggs that were small, abnormally appearing and unfertile (hereafter termed "small eggs"; Fig. 1).

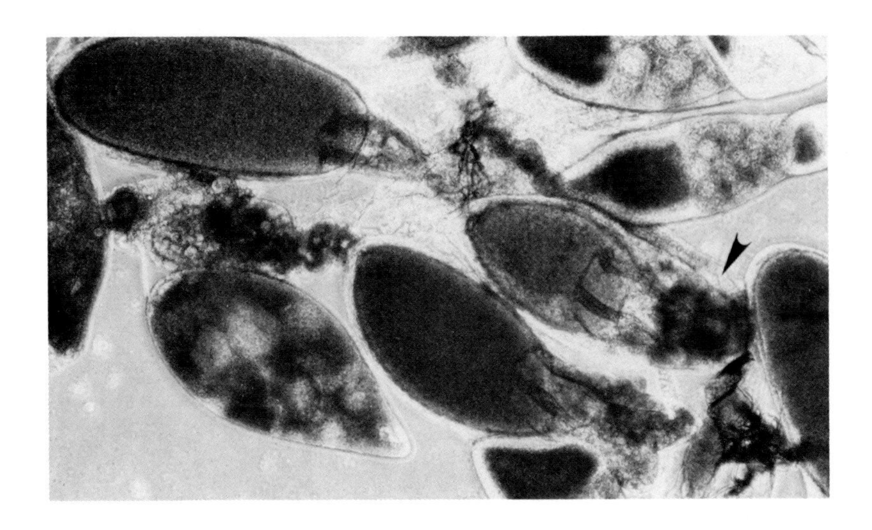

FIGURE 1. **The "small egg" phenotype of Drosophila antisense flies.** A fly carrying an integrated antisense ribosomal protein rpA1 gene driven by an heat shock promoter was given a 1-hour heat treatment. One day later the ovary was dissected and photographed. The two eggs at the center of the photograph are abnormally small while the egg at the upper left has close to normal dimensions . The arrowhead points to a cluster of nurse cells attached to a very small egg. Presumably the small size of the egg is due to failure of the nurse cells to transfer their contents to the oocyte at the end of oogenesis.

The effect was clearly dependent on the expression of the antisense genes. When flies were reared at 18°C they produced only 1-2% small eggs but at 29°C more than half of the eggs were abnormal. This difference is presumably due to the leakiness of the heat shock promoter at higher temperatures. Control flies did not lay significant numbers of small eggs at any temperature. Production of small eggs was also dependent on time after heat treatment. After a 1-hour heat pulse at 37°C, most of the eggs laid during the first day were normal; the proportion of small eggs then gradually increased to reach up to 80% on the second day and then gradually returned to normal. This observation suggests that certain stages of oogenesis (possibly vitellogenic stages 8 to 10) are more affected than others. The severity of the phenotype was also dependent on gene dosage. When the antisense genes from two independent lines were genetically combined into a single fly, the females had only rudimentary ovaries and were almost completely sterile when reared at 25°C. Male fertility appeared not to be affected in antisense flies.

Expression of Antisense rpAl RNA in Transformed Flies

The expression of the transformed genes in the transgenic flies was assessed by RNA blot analysis. The blots were hybridized with a double-stranded probe thus allowing for the simultaneous detection of the transcripts from the transformed and from the endogenous genes (these two classes of RNAs have different sizes). The RNAs from the transformed genes were undetectable in flies kept at 25°C but accumulated rapidly to several-fold over the endogenous RNA level in the antisense and in the sense flies after a 1-hour treatment at 37°C. The severity of the phenotype correlated well with the level of antisense RNA expression.

In larvae and pupae as much antisense RNA accumulated after a 1-hour heat pulse as it did in ovaries after identical treatment. Thus, the difference in phenotypic response between oogenesis and post-embryonic stages is not due to a difference in antisense induction or accumulation.

In an attempt to determine how the antisense RNA exerts its effects, we assayed for the formation of double-stranded RNA. Antisense gene expression was induced, the ovaries or whole flies were homogenized, and the RNA was extracted. Either the homogenate or the extracted RNA was digested with varying amounts of RNAase A or T1, and the resistant RNA was analyzed by electrophoresis, blotting and hybridization with a rpAl probe. In no case was double-stranded RNA detected. When a control double-stranded RNA synthesized *in vitro* was mixed with the homogenate and analyzed in the same way, a protected double-stranded RNA fragment was detected even at the highest RNAase concentration. These results suggest that double-stranded RNA either does not form or is very unstable *in vivo*, or that the double-stranded region is too small to be detected by the techniques employed.

Flies carrying two antisense genes per haploid genome ("doubly transformed flies") expressed antisense RNA at unusually high levels (10- to 20-fold the level of singly transformed flies). The level of rpAl and other mRNAs in these flies was measured by Northern blot analysis before and after induction of the antisense rpAl genes. Relative to ribosomal RNA, the level of rpAl mRNA decreased substantially after antisense gene expression. Interestingly, the levels of mRNAs coding for actin, for tubulin and for the rp49 ribosomal protein were also reduced in the doubly transformed flies. We hypothesize that in the presence of antisense rpAl RNA, synthesis of rpAl protein is decreased. As a consequence, defective ribosomes that lack this essential ribosomal protein would accumulate. These defective ribosomes would interfere with the translation of cellular mRNAs, rendering them unstable. Although we have no evidence to support this hypothesis, we hope that future experiments will confirm it.

Why are small eggs formed after antisense gene expression?

The small egg phenotype could possibly arise from insufficient synthesis of the egg components. However, close examination of the ovaries from the antisense flies suggests a different mechanism. In wild type Drosophila the vast majority of the egg constituents are synthesized by the nurse cells and transferred to the oocyte at the end of oogenesis via cytoplasmic bridges. After transfer, the follicle cells secrete the egg shell around the oocyte. In antisense flies one frequently observes mature stage 14 oocytes with a "cap" of nurse cells at their anterior pole (Fig. 1). These nurse cells obviously failed to transfer their contents to the oocyte. We interpret these observations to mean that antisense gene expression delays the developmental program of nurse cells more than that of follicle cells. Possibly, the follicle cells secrete the egg shell ahead of the nurse cell's schedule, sealing off the oocyte from the nurse cells. This would result in incomplete transfer and in the small egg phenotype.

DISCUSSION

The introduction of an antisense ribosomal protein gene into the fly's genome produced a conditional female-sterile phenotype. Sterility was obtained either by growth at elevated temperature as for conventional temperature-sensitive mutants or by a brief exposure to heat. The phenotype is strictly dependent on the expression of the antisense gene. High doses produce a more severe phenotype. Insertional mutagenesis by the P-element can be ruled out as the cause of the defect since the same phenotype was observed in three independent transformant lines. The exquisite sensitivity of oogenesis to antisense gene expression may be related to the extremely high rates of ribosome and protein synthesis during oogenesis. During the vitellogenic stages surplus ribosomal protein may be very low or not exist at all, explaining why even a modest reduction in the rate of ribosome of protein synthesis has such a profound effect on the development of the egg chamber. A similar small egg phenotype was observed by Li (4) in a Minute mutant and again 30 years later by Farnsworth (5). Despite this similarity of the oocyte phenotype between antisense and Minute flies, it is not clear why the antisense flies did not exhibit the other characteristic Minute phenotypes such as delayed development and small, thin bristles. We hypothesize that germ cells are much more sensitive to the effects of antisense gene expression than are somatic cells. The fact that increased antisense gene expression in the doubly transformed flies led to female sterility is consistent with this hypothesis. That the developemental program of germ line nurse cells may be delayed relative to somatic follicle cells (cf. previous section) is also in agreement with the hypothesis.

The disruption of cellular functions by antisense genes has previously been reported in bacteria, Dictyostelium and tissue culture cells. In Drosophila, the only experiments that have been reported relied on injection of a large excess of antisense RNA into early embryos. The present experiments are to our knowledge, the first experiments with an integrated antisense gene in a metazoan. A similar approach may be useful for the study of gene function in higher organisms (including mammals), for which transformation of cloned genes is feasible but conventional genetic screening is not practical.

REFERENCES

1. Kay MA, Jacobs-Lorena M (1987). Developmental genetics of ribosome synthesis in Drosophila. Trends in Genetics 3:347.
2. Kongsuwan K, Yu Q, Vincent A, Frisardi MC, Rosbash M, Lengyel JA, and Merriam J (1985). A Drosophila *Minute* gene encodes a ribosomal protein. Nature 317:555.
3. Qian S, Zhang J-Y, Kay MA, Jacobs-Lorena M (1987). Structural analysis of the Drosophila rpA1 gene, a member of the eucaryotic "A" type ribosomal protein family. Nucleic Acids Res 15:987.
4. Li J-C (1927). The effect of chromosome aberrations on development in *Drosophila melanogaster*. Genetics 12:1.
5. Farnsworth MW (1957). Effects of homozygous first, second and third chromosome *Minutes* on the development of *Drosophila melanogaster*. Genetics 42:19.

Stress-Induced Proteins, pages 137–148

MOLECULAR EVENTS IN AVIAN CELLS STRESSED BY HEAT SHOCK AND ARSENITE[1]

by

Milton J. Schlesinger, Nancy C. Collier,
Neus Agell, and Ursula Bond[2]
Department of Microbiology and Immunology
Washington University School of Medicine
St. Louis, MO 63110

ABSTRACT In addition to the induction of heat shock proteins, primary cultures of chicken embryo fibroblasts that have been stressed by a temperature (45°C, 1hr) or chemical agent (100μM arsenite) show a variety of changes in their metabolic activity and their morphology. Under our conditions of stress, protein synthetic capacity decreases very little (<20%) whereas DNA synthesis virtually stops and total RNA synthesis decreases by >50%. Processing of rRNA and splicing of mRNA are inhibited with prolonged stress. The free ubiquitin pool decreases slightly with a concomittant increase in the level of ubiquitin conjugates; however, one specific ubiquitin conjugate, U-H2A, loses its ubiquitin almost immediately after the stress begins. Both ubiquitin pools rise during prolonged stress but turnover of protein does not increase until cells are removed from stress. The intermediate filament (IF) network collapses shortly after stress is imposed. IF returns to normal after removal of stress but not if actinomycin D is present during stress, thus implicating a need for heat shock protein(s) in recovery of IF. Neither the

---

[1]This work was supported by a grant from the National Science Foundation.

[2]Present address: Department of Molecular Biophysics and Biochemistry, Yale University, New Haven, CT 06510.

intracellular pools of ATP and phosphocreatine nor the intracellular pH and calcium levels change significantly.

Antibodies specific for the chicken heat shock proteins have been used to identify some cellular activities of the heat shock proteins of Mr 70,000 and Mr 90,000. Antibodies specific for the small heat shock protein of Mr 24,000, HSP 24, have shown that cognate forms exist in normal muscle and lens. In stressed cells HSP 24 aggregates to huge perinuclear granules that appear to consist entirely of HSP 24.

INTRODUCTION AND RESULTS

Primary cultures of chicken embryo fibroblasts respond to an environmental stress such as heat shock by activating a small number of genes to produce mRNAs which are selectively translated to produce heat shock proteins. These heat shock genes and their encoded proteins are very similar in sequence and structure to genes and proteins present in virtually every organism subjected to a heat shock stress (1,2,3). Furthermore, the DNA sequences encoding the cis-acting regulatory elements in chicken heat shock genes are precisely those found in the promoter regions of all eukaryotic heat shock genes. This extreme conservation is indicative of an essential role for heat shock proteins in survival of an organism. To determine how the major heat shock proteins function in protection and recovery of stressed cells has been the goal of our research.

Our initial observation that avian and mammalian tissue culture cells induce heat shock proteins (4) was among the first indications that the heat shock response discovered much earlier by Ritossa in _Drosophila_ (5) was universal. Three major heat shock proteins were subsequently purified from stressed cultures of chicken fibroblasts and used to raise monospecific, polyclonal antibodies in rabbits (6). The antibodies directed against the two higher molecular weight proteins showed cross reactivity with heat shock proteins of similar molecular weight from species as diverse as yeast and human and they have proved useful in identifying several roles for the HSP 70 and HSP 90. These include the initial identification of the bovine clathrin uncoating

ATPase as a member of the mammalian HSP 70 protein "family" (7,8), the identification of the 90 kDa protein in a transient complex formed with the pp60 src phosphotyrosine protein kinase as HSP 90 (9) and the more recent discovery of the HSP 90 as a subunit of the inactive form of the murine glucocorticoid steroid receptor complex (10). Very preliminary results with our anti-HSP 70 antibodies indicate that a member of the HSP 70 family is part of a complex required for the ATP-dependent import of nuclear-encoded proteins into yeast mitochondria (K. Verner and G. Schatz, personal communication). The latter would be consistent with recent data from other laboratories (see papers by E. Craig et al. and G. Waters et al. in this volume) in which an HSP 70 protein is a component of the system for importing folded proteins into the endoplasmic reticulum.

All of the data accumulated thus far from many laboratories indicate that HSP 70 family members form complexes with various cellular proteins and -in the presence of ATP- dissociate these complexes by altering the conformation of the proteins in the complex (11). The HSP 90 also forms dissociable complexes- but the factors required for dissociation bind to sites not on the HSP 90 but on those polypeptides to which the HSP 90 is bound. In the process of purifying the major chicken heat shock proteins, we found a large fraction of these proteins as high molecular weight oligomers (6). We have not identified any protein in the chicken fibroblast that is complexed with HSP 70 or HSP 90 although we have noted an ATP-dependent release of HSP 70 from the nuclear cytoskeleton fraction of heat-shocked fibroblasts (12)- an activity previously reported for mammalian cells (13).

A role for the small HSP's with mol.wts. of 20-30 k has not been found. We have devoted a considerable effort over the past ten years to the purification of HSP 24 and investigation of its function in the chicken cell. In earlier studies, we found almost all of the HSP 24 formed after a 4-hour heat shock in the cytosol fraction of cell extracts and about half of the total HSP 24 eluted from a Sepharose 6B column with a native molecular weight of 180,000 (6). The balance of the protein was in higher molecular weight complexes. Attempts to purify these complexes by conventional protein purification procedures invariably led to losses by aggregation to insoluble forms. Insoluble forms of HSP 24 were also found in cells after long periods of stress or after a second stress imposed after a 12 hour recovery period. The HSP 24 remained in an insoluble form after extraction of cells and could only be solubilized by strong denaturants such as ionic detergents, urea and guanidine. Under these same conditions of extended stress, phase-dense granules appeared. We utilized our rabbit antibodies raised against purified HSP 24 to show that these heat shock granules contained this small heat shock protein (12). High resolution electron micrographs of thin sections showed the granules to be dense proteinaceous meshworks (14). This structure was also seen in replicas of cells that were quick frozen, deep etched and freeze fractured (Fig.1). Antibodies were also used in the ultrastructural studies to confirm that the structures visualized contained the HSP 24. We were unable to detect RNA in the fibrous structure and crosslinking experiments failed to detect any other protein except the HSP 24. Thus, we conclude that the small heat shock protein can- under appropriate conditions- aggregate with itself to form huge polymeric structures. In the cell, this polymer is freely reversible- the granule "dissolves" and HSP 24

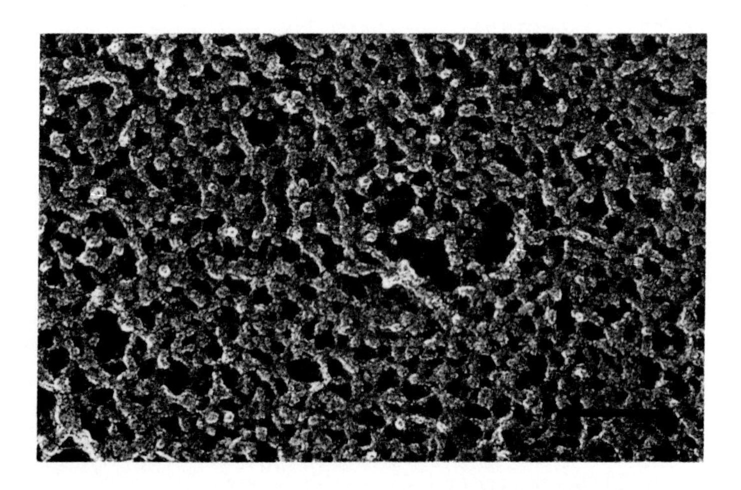

FIGURE 1. Ultrastructure of the HSP 24 heat shock granule. Restressed cells were treated with hypotonic buffer prior to freezing. Preparations were freeze-fractured and replicas prepared according to the procedure of Heuser (15). Bar 0.1 um.

becomes "soluble" upon recovery from the extended stress; however, we have been unable to find conditions other than denaturants that will solubilize the complex in vitro.

Our initial studies with anti-HSP 24 antibodies revealed a cross-reactive protein of mol. wt. 22-24,000 present in all muscles tissues of the embryonic and adult chicken in the absence of any stress. More recent experiments with a preparation of antibodies affinity-purified with the heat shock protein confirm the presence of this "cognate" form which is also found in the lens (16). The lens cognate has a mobility in SDS/PAGE identical to the chicken beta crystallin. Because of this we had erroneously reported that HSP 24 is immunologically related to the beta crystallin (17), but we subsequently separated the cognate from the beta crystallins. We employed immunohistochemical staining on skeletal muscle tissue sections to localize the HSP 24 cognate and found that this protein is present in muscle rather than in surrounding cells and is not associated with any of the established organelles or structures in muscle tissue. In Western blots probed with anti HSP 24 antibodies, we detect higher molecular weight forms of a cross-reacting protein in extracts of primary muscle cultures and in embryonic chicken brain (Fig.2). These forms do not appear to be precursors to the 22,000 mol. wt. cognate. We do not know if the cognate form is developmentally expressed in the chicken in the manner shown for the small heat shock proteins of Drosophila. The only properties of the chicken HSP 24 that are related to small heat shock proteins from other organisms are the reversible formation in vivo of high molecular weight aggregates and its amino acid composition which is similar to compositions determined from the amino acid sequences derived from cDNA sequences (14). The propensity for aggregation probably results from the hydrophobic character of the native form of soluble HSP 24. We demonstrated the latter by showing that a significant fraction of chicken HSP 24 partitioned into the nonpolar layer of a Triton X-114 biphasic solution. Based on all of our data, we postulate that the polymeric state of HSP 24 could represent an allosteric and highly stable form of an enzyme used by cells to recover from stress.

The formation of heat shock proteins is clearly only one of many events occurring in a stressed cell. What are some of the others? We have addressed this question over

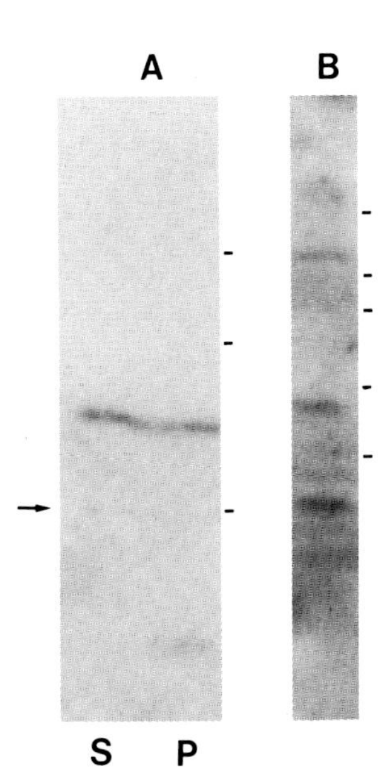

FIGURE 2. A. Immunoblot analysis of embryonic brain. Brains from 11-day chicken embryos were washed with 50 mM Tris-Cl, pH 7.5, 100 mM NaCl, 1 mM EDTA, 5 mM PMSF and sonicated in the same buffer. Insoluble material was pelleted in a microfuge and both the supernatant and pellet fractions were boiled with Laemmli buffer. Proteins (about 100 ug) were separated by sDS/PAGE on a 12.5% acrylamide gel and electrophoretically transfered to nitrocellulose. The blot was probed with anti-HSP 24 antibody and ^{125}Iodinated protein A. The antibody recognized small but detectable band at Mr 24,000 (arrow). Molecular weight markers are denoted by bars and correspond to Mr 70,000, 43,000 and 24,000. B. Immunoblot analysis of muscle cultures. Five-day old cultures of pectoral muscle with the majority of cells exhibiting multi-nucleation were extracted in Laemmli sample buffer. Samples were separated by SDS/PAGE in a 11% acrylamide gel and analyzed as described in A. Molecular weight markers are denoted by bars and correspond to Mr 84,000, 58,000, 49,000, 37,000 and 27,000.

the past years during our study of the heat shocked chicken fibroblast cell culture system and summarize here results of those varied studies. First, however, we consider it crucial to define the stress we use- for it is clear that the extent of stress experienced by a cell or tissue system can have profound effects on the metabolic state of the organism. We employ mainly two kinds of stress: (1) a heat shock of 45° which is within the normal physiological range of the organism and (2) incubation with 100 μM sodium arsenite. For periods up to 4 hours for the former and 2 hours for the latter, there is less than 20% loss in the protein synthetic capacity of the cell and the pattern of protein revealed by one dimensional SDS/PAGE shows very little loss of normal cell proteins but the addition of major heat shock proteins. The latter accounts for about 15% of the total protein systhesis measured after a one hour stress (18).

When stressed by the conditions noted above, primary cultures of chicken fibroblasts show a dramatic decline in DNA synthesis (measured by ^{3}H-thymidine uptake) almost immediately after the stress. After 1 hour of stress ^{3}H-thymidine incorporation into trichloroacetic acid precipitable material is inhibited by 80-90% of that of the non-stressed cell. In contrast, ^{3}H-uridine uptake is blocked by about 50%. Fatty acid incorporation into lipid as measured by ^{3}H-palmitic acid was reduced by 20%. Incorporation of ^{3}H-mannose into protein was not significantly inhibited. We have shown elsewhere that these stressed cells have defects in post-transcriptional processing of mRNAs leading to formation of mRNAs with introns (19). We have also seen a block in processing of ribosomal RNAs (P. Kelley, Ph.D. thesis, Washington U., 1979). In contrast, we find little effect on post translational processing or transport of secreted glycoproteins as measured by the pattern of ^{3}H-mannose label in proteins separated by SDS/PAGE. Others have reported changes in phosphorylation and dephosphorylation of cytoplasmic and nuclear proteins (noted in 20). We have not examined this kind of post translational modification in our cells under our conditions of stress.

We expected that one particular activity in the cell would be affected by stress--namely the turnover of protein since it has been widely assumed that most stress agents lead to denaturation of proteins. In our studies, we find little change in protein turnover <u>during</u> the period of stress, but turnover rates double immediately upon removing cells from the stress (18). The major mechanism for removing denatured proteins involves the ubiquitin-dependent proteolysis pathway in which proteins targeted for degradation are conjugated with multiple ubiquitins. Both ubiquitination and subsequent proteolysis are ATP dependent reactions. We earlier showed that ubiquitin is a heat shock protein and the ubiquitin gene which codes for multiple copies of ubiquitin contains the conserved heat shock promoter in its 5' untranscribed sequence (19). Shortly after stress free ubiquitin decreases by 20% with a concomitant rise in ubiquitin conjugates. Later both free and conjugated ubiquitin levels rise as levels of polyubiquitin increase. These data are consistent with a demand in the cell for more ubiquitin to handle increased amounts of denatured proteins, but it is unclear why the cells wait until after the stress to degrade these ubiquitin conjugates. It is not because of a limiting amount of ATP since levels of ATP and phosphocreatine do not change significantly in the stressed cell (Table 1). In sharp contrast to the increased amount of ubiquitin conjugates formed after heat shock, the ubiquitinated histone H2A is rapidly deubiquitinated after stress (18).

TABLE 1
LEVELS OF ATP AND PHOSPHOCREATINE (PCr) IN ARSENITE STRESSED CELLS

Time of Stress	ATP	PCr
0	10.1	10.3
5 min.	9.6	8.9
20 min.	9.4	8.8
60 min.	9.4	9.4
60 min. + 10 min. recov.	10.1	9.1
60 min. + 60 min. recov.	8.0	9.1

Values are nmoles per 10^6 cells

One of the more dramatic changes in the stress fibroblast is a collapse of the intermediate filament network (12). Under the same stress conditions, no changes are detected in the microfilament and microtubule network or in the nuclear lamin morphology. Very shortly upon removal of the stress, the intermediate filament returns to its extended morphology. However, this recovery does not occur if cells are treated with actinomycin D during stress. Based on this result, we have hypothesized that a heat shock protein is needed to restore the intermediate filament network.

Several other measurements have been made on the stressed chicken cells in order to look for changes in general metabolic activities. Stressed chicken cells appear to shift to glycolysis as source for ATP generation- lactic acid formation and glucose uptake increases significantly. In several organisms, glycolytic enzymes increase after heat shock (21-23). However, in contrast to reports on other stressed cells (24,25), chicken cells show no significant change in intracellular levels of calcium or pH after arsenite stress. We also find no change in levels of inositol trisphosphate indicating that our stress conditions do not perturb the phosphotidylinositol pathway which acts to trigger a variety of cellular activities upon stimulation by various growth factors interacting at the cell surface.

Based on the data we have accumulated thus far, the two most sensitive responders in the stressed fibroblast culture are nuclear metabolism and the intermediate filament network. It is possible that there is, in fact, a relationship between these for it has been suggested that the intermediate filament network "communicates" events in the cell cytoplasm and membranes to the nucleus. We interpret the nuclear events - the stop in DNA synthesis and histone deubiquitination- as indicative of the cell slowing its overall growth during stress. Even before these cellular responses occur the heat shock genes are probably activated to transcribe mRNAs. Ritossa's data on the Drosophila polytene chromosome showed puffing was a very rapid response. The recent data of Lis (noted in a chapter here) show the presence of RNA polymerase II on the Drosophila HSP 70 gene and its initiation of transcription prior to a heat shock. Thus, the signal for heat shock mRNA transcription probably is distinct from those leading to the varied events described here. These two distinctive events - turn on of heat shock genes and turn off of several cellular activities - appear to constitute the "heat shock response". How do heat shock proteins fit into this picture? We believe that ubiquitin provides a paradigm for the role of heat shock proteins in the overall stress response in the following way. First, a normal ubiquitin conjugate - histone H2A - is altered in a way that ultimately leads to the slowing of DNA replication and transcription. Second, misfolded proteins become ubiquitinated in order to be eliminated from the cell. Third, the polyubiquitin gene is activated to produce more ubiquitin. Fourth, the recovery of the cell is accompanied by elimination of the ubiquitin conjugates and the restoration of ubiquitinated histones (note that the later has not been directly demonstrated). The final outcome is a protection of the stressed cell through deubiquitination of histones and ubiquitination of denatured protein followed by an ability for the cell to return to its previous metabolic state through elimination of ubiquitinated proteins and the employment of the newly made ubiquitin for reformation of the histone conjugate.

ACKNOWLEDGMENTS

We thank Dr. John Heuser for his assistance with the ultrastructure studies, Dr. Oliver Lowry for measurements of ATP and phosphocreatine levels, Dr. Paul Schlesinger for measurements of the intra-cellular pH and calcium concentrations and Dr. Phillip Majerus for inositol levels.

REFERENCES

1. Morimoto RI, Hunt C, Huang S-Y, Berg KL, Banerji SS (1986). Organization, nucleotide sequence, and transcription of the chicken HSP 70 gene. J Biol Chem 261:12692.
2. Binart N, Chambrand B, Dumas B, Garnier J, Baulieu EE (1988). Cloning and nucleotide sequence of the chick 90 kDa heat shock protein. J. Cell Biochem 12 D:275.
3. Bond U, Schlesinger MJ (1985). Ubiquitin is a heat shock protein in chicken embryo fibroblasts. Mol Cell Biol 5:949.
4. Kelley PM, Schlesinger MJ (1978). The effect of amino acid analogues and heat shock on gene expression in chicken embryo fibroblasts. Cell 15:1277.
5. Ritossa FM (1982). A new puffing pattern induced by heat shock and DNP in Drosphila. Experienta 18:571.
6. Kelly PM, Schlesinger MJ (1982). Antibodies to two major chicken heat shock proteins crossreact with similar proteins in widely divergent species. Mol Cell Biol 2: 267.
7. Chappell TG, Welch WJ, Schlossman DM, Palter KB, Schlesinger MJ, Rothman JE (1986). Uncoating ATPase is a member of the 70 kDa family of stress proteins. Cell 45:3.
8. Schlesinger MJ (1986). Heat shock proteins: the search for functions. J Cell Biol 103:321.

9. Opperman H, Levinson W, Bishop, JM (1981). A cellular protein that associates with a transforming protein of Rous Sarcoma virus is also a heat shock protein. Proc Natl Acad Sci USA 78:1067.
10. Sanchez EH, Toft DO, Schlesinger MJ, Pratt WB (1985). Evidence that the 90 kDa phosphoprotein associated with the untransformed L-cell glucocorticoid receptor is a murine heat shock protein. J. Biol Chem 260:12398.
11. Pelham H (1985). Activation of heat shock genes in eukaryotes. Trends in Genetics 1:21.
12. Collier NC, Schlesinger MJ (1986). The dynamic state of heat shock proteins in chicken embryo fibroblasts. J Cell Biol 103:1495.
13. Pelham HRB (1984). Hsp 70 accelerates the recovery of nucleolar morphology after heat shock. EMBO J 3:3095.
14. Collier NC, Heuser J, Levy MA, Schlesinger MJ (1988). Ultrastructural and biochemical analysis of the stress granule in chicken embryo fibroblasts. J Cell Biol 106:1131.
15. Heuser JE (1980). Three-dimensional visualization of coated vesicle formation in fibroblasts. J Cell Biol 84:560.
16. Collier NC, Schlesinger MJ (1986). Induction of heat shock proteins in the embryonic chicken lens. Exp Eye Res 43:103.
17. Schlesinger MJ (1985). Stress response in avian cells. In Atkinson BG, Walden DG (eds): "Changes in eukaryotic gene expression in response to environmental stress" Orlando, Florida: Academic Press, p.183.
18. Bond U, Agell N, Haas AL, Redman K, Schlesinger MJ (1988). Ubiquitin in stressed chicken embryo fibroblasts. J Biol Chem 263:2384.
19. Bond U, Schlesinger MJ (1986). The chicken ubiquitin gene contains a heat shock promoter and expresses an unstable mRNA in heat shocked cells. Mol Cell Biol 6:4602.
20. Schlesinger MJ, Ashburner M, Tissieres A (eds) (1982). "Heat Shock from Bacteria to Man". New York: Cold Spring Harbor Laboratory.
21. Hammond GL, Lai Y-K, Markert CL (1982). Diverse forms of stress lead to new patterns of gene expression through a common and essential metabolic pathway. Proc Natl Acad Sci USA 79:3485.
22. Browder LW, Pollack M, Nickells RW, Heikkila JJ, Winning RS (1988). Developmental regulation of the

heat shock response. in DiBerardino MA, Etkin LD (eds) Developmental Biology: A comprehensive synthesis (in press).

23. Iida H, Yahara I (1985). Yeast heat shock protein of Mr 48,000 is an isoprotein of enolase. Nature 315:688.
24. Stevenson MA, Calderwood SK, Hahn GM (1986). Rapid increases in inositol trisphosphate and intracellular Ca^{2+} after heat shock. Biochem Biophys Res Commun 137:826.
25. Drummond IAS, McClure SA, Poenie M, Tsien RY, Steinhart RA (1986). Large changes in intracellular pH and calcium observed during heat shock are not responsible for the induction of heat shock proteins in Drosophila melanogaster. Mol Cell Biol 6:1767.

Stress-Induced Proteins, pages 149–159

UBIQUITIN METABOLISM IN STRESSED MAMMALIAN CELLS

G. Pratt, Q. Deveraux, and M. Rechsteiner[1]

Department of Biochemistry
University of Utah School of Medicine
Salt Lake City, Utah 84132

ABSTRACT HeLa cells were injected with trace amounts of radioiodinated ubiquitin and then exposed to arsenite, ethanol or elevated temperature. In response to each metabolic stress, the cells increased their levels of high molecular weight ubiquitin conjugates and decreased their content of ubiquitinated histones. These alterations in ubiquitin pools were extremely rapid. After warming the cells to 45°C the half-time for accumulation of high molecular weight conjugates was approximately 5 minutes, and complete equilibration was reached within 20 minutes. In contrast, when HeLa cells were fused to red blood cells loaded with large amounts of ubiquitin and then placed at 45°C, there was little or no change in the distribution of ubiquitin among HeLa proteins. Our ability to suppress shifts in ubiquitin pools by increasing the intracellular concentration of ubiquitin should allow us to determine whether changes in ubiquitin distribution are essential to the heat shock response.

INTRODUCTION

It has been suggested that biological macromolecules arose several billion years ago during cycles of heating/evaporation and rehydration. If true, the heat shock response must have been one of the first

---

[1]The studies presented here were supported by N.I.H. grant GM27159.

homeostatic mechanisms to appear in primitive cells. An early origin of the response is certainly suggested by its universal occurrence and the impressive homology of several procaryotic and eucaryotic heat shock proteins (1,2). Until recently, the function of heat shock proteins (HSPs) has been a mystery. Now, however, it has become clear that some HSPs are involved in the recognition of unfolded, misfolded or unassembled polypeptides. It has also become increasingly evident that even in the absence of metabolic stress, cells express proteins similar or identical to HSPs. For example, HSP70 and HSP90 have been implicated in the transit of newly-synthesized proteins to their proper cellular compartments (3-5). In _E. coli_, the heat-inducible, protease La contributes to intracellular proteolysis at low temperatures as well (6).

These two emerging attributes of HSPs, recognition of aberrant protein structures and expression under normal physiological conditions, are well illustrated by ubiquitin. This remarkable little eucaryotic protein, which is an abundant constituent of eucaryotic cells under all growth conditions, plays a key role in intracellular proteolysis (7). Ubiquitin can be covalently attached to other cellular proteins,and this post-translational modification is thought to target proteins for destruction.

There have also been proposals that ubiquitin may help to regulate the heat shock response in eucaryotes. Munro and Pelham (8) have suggested that under normal conditions, heat shock transcription factor(s) is normally inactivated by ubiquitination and activated when stress-induced generation of abnormal proteins produces high levels of substrates for ubiquitin attachment. A similar proposition has been advanced by Finley _et al_. (9). Both hypotheses predict changes in the spectrum of ubiquitinated proteins following heat shock. This was tested by introducing labeled ubiquitin into cultured cells.

MICROINJECTION OF UBIQUITIN

Some years ago we developed a method for injecting proteins into large numbers of cultured mammalian cells (10). Red blood cells are lysed and resealed in the presence of exogenous proteins, and during lysis some of the exogenous protein becomes entrapped within the red cells. These "loaded" red cells are then fused to cultured cells using Sendai virus or polyethylene glycol. Recently, we used this methodology to examine ubiquitin pool dynamics in stressed and non-stressed HeLa cells (11,12).

Radioiodinated ubiquitin was introduced into HeLa cells by erythrocyte-mediated microinjection. Subsequent electrophoretic analyses revealed that the injected ubiquitin molecules were rapidly conjugated to HeLa proteins. At equilibrium, 10% of the injected ubiquitin was covalently attached to histones, and 40% was distributed among conjugates of higher molecular weight. Although the remaining ubiquitin molecules appeared to be unconjugated, the free pool of ubiquitin decreased by one-third and additional conjugates were present when electrophoresis was performed at low temperature under nonreducing conditions. Molecular weights of these labile conjugates suggest that they are ubiquitin adducts in thiolester linkage to activating enzymes. Despite the fairly rapid degradation of injected ubiquitin ($t_{1/2} \sim 10$-20 h), the size distribution of ubiquitin conjugates within interphase HeLa cells remained constant for at least 24 h after injection.

However, when the injected cells were incubated at 45°C for 5 minutes (reversible heat-shock) or for 30 minutes (lethal heat-shock), there were dramatic changes in the levels of ubiquitin conjugates (12). The free ubiquitin pool and the level of histone-ubiquitin conjugates decreased rapidly, and high molecular weight conjugates predominated. Formation of large conjugates did not require protein synthesis. When analyzed by two-dimensional electrophoresis, the major conjugates did not co-migrate with heat-shock proteins before or after thermal stress.

Concomitant with the loss of free ubiquitin, the degradation of endogenous proteins, injected hemoglobin, BSA, and ubiquitin was reduced in heat-shocked HeLa cells. After reversible heat-shock, the decrease in proteolysis was small, and both the rate of proteolysis and the size of the free ubiquitin pool returned to pre-stress levels upon incubation at 37°C. In contrast, neither proteolysis nor free ubiquitin pools returned to control levels after lethal heat-shock. Nevertheless, lethally heat-shocked cells degraded denatured hemoglobin more rapidly than native hemoglobin, and ubiquitin-globin conjugates formed within them. Therefore, stabilization of proteins after heat-shock was not due to the loss of ubiquitin conjugation or inability to degrade proteins that form conjugates with ubiquitin.

Since treatments other than heat are known to induce synthesis of stress proteins (13,14), we have asked whether chemical agents also affect the size distribution of intracellular ubiquitin conjugates. HeLa cells were injected with labeled-ubiquitin and then placed at 45°C or exposed to arsenite or ethanol. At various times thereafter, cells were dissolved in sample buffer, and the size distribution of ubiquitin conjugates was determined by SDS-PAGE. It can be seen from the data in Figure 1 that arsenite and ethanol, like heat, cause high molecular weight conjugates to accumulate and ubiquitinated histones to diminish. Responses of the ubiquitin pools to chemical stresses were, however, considerably slower than that observed after shift to 45°C. The rapid response of the ubiquitin pools to high temperature is quite striking and re-emphasizes the dynamic nature of ubiquitination reactions.

CONJUGATE POOL DYNAMICS IN CELLS CONTAINING EXCESS FREE UBIQUITIN

As noted above, two groups of investigators have proposed that ubiquitin might play a central role in regulating the heat shock response (8,9). Both hypotheses are based on the idea that HSP transcription factors are activated (or spared) when the intracellular pool of ubiquitin is depleted by its ligation to denatured proteins. Our earlier demonstration that heat shock produces substantial changes in the distribution of ubiquitin (12) plus the data in Figure 1 showing that arsenite and ethanol also affect conjugate patterns are consistent with such a possibility. If ubiquitin depletion were the signal for increased HSP transcription, then artificially increasing the concentration of free ubiquitin might prevent any significant decrease in ubiquitin and could inhibit the heat shock response. With this in mind, we fused HeLa cells to red blood cells osmotically lysed in the presence of 40 mg/ml ubiquitin plus trace amounts of the radioiodinated molecule. As shown in Figure 2, the steady state distribution of ubiquitin was clearly altered by the high concentration of

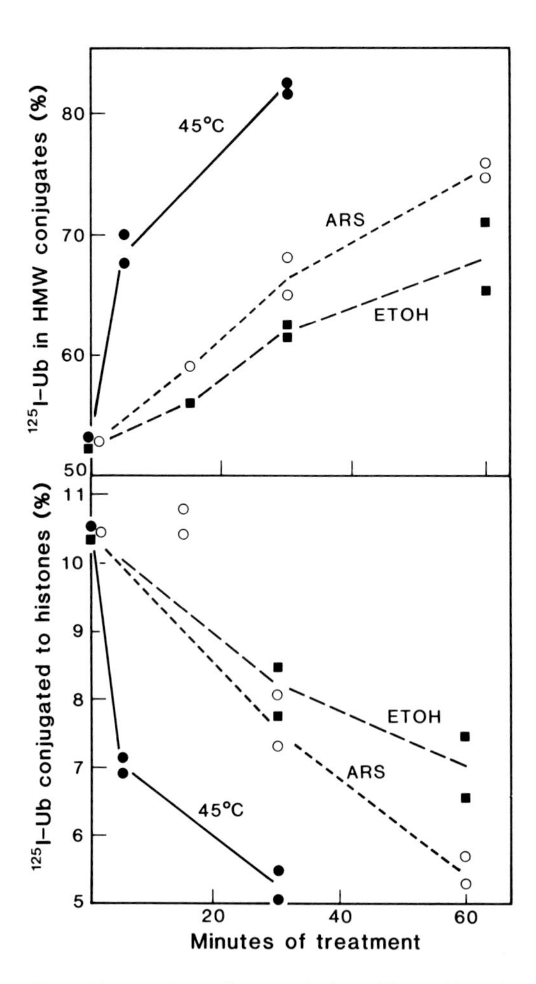

Figure 1. The molecular weight distribution of ubiquitin conjugates in HeLa cells subjected to stress.

HeLa cells were microinjected with radioiodinated ubiquitin, and six hours later the cells were warmed to 45°C or placed in medium containing 100 μM sodium arsenite or 6% ethanol. Molecular weight distributions of ubiquitin conjugates were determined as previously described (11,12). The upper panel shows the rate at which high molecular weight (>25 kDa) conjugates accumulate; the lower panel shows the kinetics of uH2A disappearance.

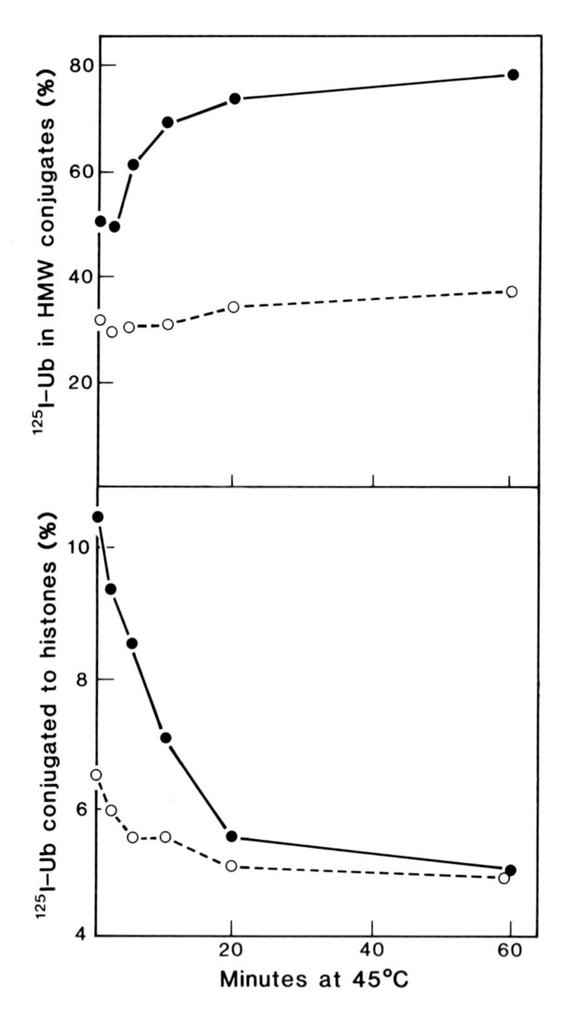

Figure 2. Ubiquitin pool dynamics in HeLa cells containing excess free ubiquitin.

HeLa cells were injected with trace amounts of radioiodinated ubiquitin alone (●) or labeled ubiquitin diluted with 40 mg/ml unlabeled ubiquitin (o). Samples were taken at various times after shifting the cells to 45°C, and the size distribution of ubiquitin conjugates was determined as described in the legend to Figure 1.

unlabeled molecules. Whereas iodinated ubiquitin usually partitions 50% to high molecular weight conjugates, 10% to uH2A/uH2B and 40% the "free" pool, the distribution was 30% in high molecular weight conjugates, 6% in histones, and 65% of the molecules were present in the "free" pool in HeLa cells receiving excess ubiquitin. Moreover, after transferring each set of cells to 45°C, there were substantial shifts in the ubiquitin conjugates of HeLa cells containing normal ubiquitin levels, but almost no change was observed in HeLa cells injected with large amounts of ubiquitin (see Figure 2). We are now asking whether excess ubiquitin can prevent synthesis of HSPs after stress. Unfortunately, those experiments have not been completed.

DO DENATURED PROTEINS SIGNAL THE HEAT SHOCK RESPONSE?

In 1980, Hightower suggested that HSPs are synthesized in response to abnormal proteins and that HSPs maybe involved in the degradation of these newly formed substrates. In support of this hypothesis it has been found that HSPs are synthesized in *Drosophila* cells producing mutant actins (15), injection of denatured proteins into oocytes can induce HSP synthesis (16) and agents that stabilize proteins can inhibit certain inducers of the heat shock response (17). The results presented above are also consistent with the hypothesis. Ubiquitin is preferentially ligated to denatured or abnormal proteins (18,19), so the observed increase in high molecular weight conjugates at 45°C can be rationalized in terms of protein denaturation within stressed cells.

Recent studies on ubiquitin metabolism in the mouse lymphoma cell line, ts85, indicate that such an interpretation may be too simple. Ts85 cells are known to have a temperature-sensitive lesion in the ubiquitin-activating enzyme, E1; they also express HSPs at 39°C, the non-permissive temperature (9). Using western blots, we examined the pattern of ubiquitin conjugates in ts85 cells and the parent line, FM3A, at various temperatures. Fully expecting to see ubiquitin conjugates disappear from ts85 cells at temperatures above 39°C, we were surprised to find that high molecular weight conjugates actually accumulated in ts85 cells at 39°C and above (see Figure 3).

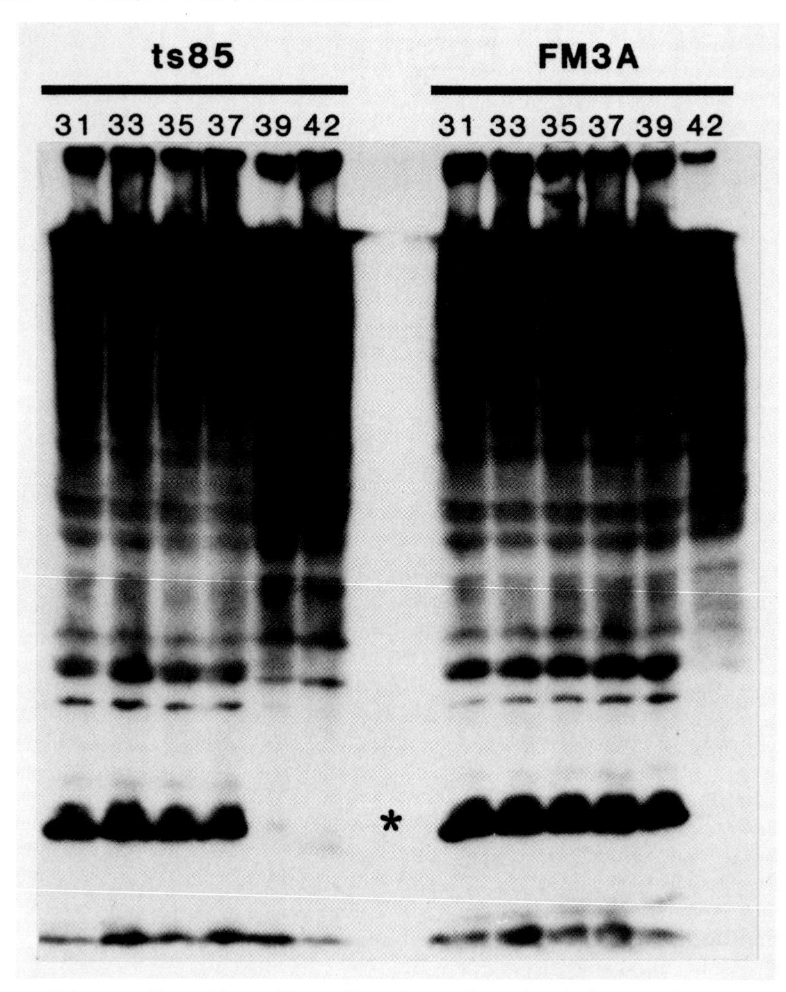

Figure 3. Size distribution of ubiquitin conjugates in ts85 and FM3A cells at various temperatures.

Either ts85 or FM3A cells were incubated for 5 hours at various temperatures between 31 and 39°C or for 1 hour at 42.5°C. After incubation the cells were dissolved in SDS-PAGE sample buffer, their proteins were separated on 10% acrylamide gels, transferred to nitrocellulose and ubiquitin conjugates were detected using rabbit anti-ubiquitin plus radioiodinated Protein A. The auto-radiogram presented above shows that whereas uH2A(*) disappears and high molecular weight conjugates accumulated in ts85 cells at 39°C or above, such changes do not occur in FM3A cells below 42°C.

By contrast, the pattern of ubiquitin conjugates did not change in FM3A cells kept below 42°C (Figure 3).

Although these results reconfirm a connection between ubiquitin metabolism and the heat shock response, the data are indeed puzzling. First, E1 is required to activate the carboxy terminus of ubiquitin, so one expects ubiquitin conjugates to disappear above 39°C. Second, if the stress-induced increase in high molecular weight conjugates reflects general protein denaturation, why should there be a noticeable increase in ts85 ubiquitin conjugates at 39°C? Presumably the lesion in ts85 is restricted to E1 and does not encompass a large number of other cellular proteins. Although it is possible that the high molecular weight conjugates actually represent ubiquitinated E1 molecules since the enzyme appears to be fairly abundant (11), this seems rather unlikely. And for the moment, we are unable to present a coherent picture of the relationship between ubiquitin metabolism and the heat shock response.

CONCLUSION

Like several other heat shock proteins, ubiquitin is part of a physiological system that responds to unassembled or abnormal protein structures. In addition, ubiquitin may play a role in regulating the stress response. The ubiquitin pool dynamics revealed by our microinjection studies are consistent with, but by no means prove, the latter possibility.

ACKNOWLEDGEMENTS

We thank Kristen Ballantyne for excellent word processing.

REFERENCES

1. Lindquist S (1986). The heat-shock response. Ann Rev Biochem 55:1151.
2. Burdon RH (1986). Heat shock and the heat shock proteins. Biochem J 240:313.
3. Deshaies RJ, Koch BD, Werner-Washburne M, Craig EA, Schekman R (1988). A subfamily of stress proteins

facilitates translocation of secretory and mitochondrial precursor polypeptides. Nature 332:800.
4. Chirico WJ, Waters MG, Blobel G (1988). 70K heat shock related proteins stimulate protein translocation into microsomes. Nature 332:805.
5. Courtneidge SA, Bishop JM (1982). Transit of $pp60^{v-src}$ to the plasma membrane. Proc Natl Acad Sci USA 79:7117.
6. Phillips TT, VanBogelen R, Neidhardt F (1984). Lon gene product of Escherichia coli is a heat-shock protein. J Bacteriol 159:283.
7. Rechsteiner M (1987). Ubiquitin-mediated pathways for intracellular proteolysis. Ann Rev Cell Biol 3:1.
8. Munro S, Pelham H (1985). What turns on heat shock genes? Nature 317:477.
9. Finley D, Ciechanover A, Varshavsky A (1984). Thermolability of ubiquitin-activating enzyme from the mammalian cell cycle mutant ts85. Cell 37:43.
10. Schlegel R, Rechsteiner M (1975). Microinjection of thymidine kinase and bovine serum albumin into mammalian cells by fusion with red blood cells. Cell 5:371.
11. Carlson N, Rechsteiner M (1987). Microinjection of ubiquitin: Intracellular distribution and metabolism in HeLa cells maintained under normal physiological conditions. J Cell Biol 104:537.
12. Carlson N, Rogers S, Rechsteiner, M (1987). Microinjection of ubiquitin: Changes in protein degradation in HeLa cells subjected to heat-shock. J Cell Biol 104:547.
13. Li GC (1983). Induction of thermotolerance and enhanced heat shock protein synthesis in Chinese hamster fibroblasts by sodium arsenite and by ethanol. J Cell Physiol 115:116.
14. Mizzen LA, Welch WJ (1988). Characterization of the thermotolerant cell. I. Effect on protein synthesis activity and the regulation of heat-shock protein 70 expression. J Cell Biol 106:1105.
15. Hiromi Y, Hotta Y (1985). Actin gene mutations in Drosophila; heat shock activation in the indirect flight muscles. EMBO J 4:1681.
16. Ananthan J, Goldberg AL, Voellmy R (1986). Abnormal proteins serve as eukaryotic stress signals and trigger the activation of heat shock genes. Science 232:522.

17. Hightower LE, Guidon PT Jr, Whelan SA, White CN (1985). In BG Atkinson, DB Walden (eds): "Changes in Eukaryotic Gene Expression in Response to Environmental Stress," New York: Academic Press, Inc., p 197.
18. Chin DT, Kuehl L, Rechsteiner M (1982). Conjugation of ubiquitin to denatured hemoglobin is proportional to the rate of hemoglobin degradation in HeLa cells. Proc Natl Acad Sci USA 79:5857.
19. Hershko A, Eytan E, Ciechanover A, Haas AL (1982). Immunochemical analysis of the turnover of ubiquitin-protein conjugates in intact cells. J Biol Chem 257:13964.

IV. FUNCTIONS OF HEAT-SHOCK PROTEINS

Stress-Induced Proteins, pages 163–174

PURIFICATION OF YEAST STRESS PROTEINS BASED ON THEIR ABILITY TO FACILITATE SECRETORY PROTEIN TRANSLOCATION[1]

M. Gerard Waters, William J. Chirico, Rubén Henríquez, and Günter Blobel

Laboratory of Cell Biology, Howard Hughes Medical Institute, The Rockefeller University, 1230 York Avenue, New York, NY 10021

ABSTRACT

Two constitutively expressed yeast stress proteins have been purified to near homogeneity based upon their ability to stimulate post-translational translocation of prepro-α-factor into yeast microsomes. The proteins are the products of the *SSA1* and *SSA2* genes. Here we describe the purification of these proteins in detail. Finally we show that purified *Escherichia coli dnaK* protein also stimulates post-translational translocation of prepro-α-factor.

INTRODUCTION

We are interested in the process of translocation of integral membrane or secretory proteins into or through the membrane of the endoplasmic reticulum. *Saccharomyces cerevisiae* was chosen as a model organism because of the availability of mutants and genetic techniques absent in higher eukaryotic systems. Several

[1]This work was supported by the Howard Hughes Medical Institute (WJC, GB) and Public Health Service Training Grant GM-07922-07 (MGW).

groups have shown that prepro-α-factor, the precursor of the yeast secretory protein α-factor, can be translocated into yeast microsomes *in vitro* in an ATP-dependent post-translational process (1-3). Recently we have developed an *in vitro* assay in which post-translational translocation of this protein is stimulated by a yeast cytoplasmic extract and have shown the active component(s) to be, at least in part, proteinaceous (4). Purification of the activity yielded a 70 kD ATP-binding protein which was resolved into two proteins of pI about 5.5 and 5.4 by two-dimensional gel electrophoresis (5). Both proteins reacted with a monoclonal antibody, raised against *Drosophila* hsp70 (antibody 7.10, reference 6), that recognizes an evolutionarily conserved epitope in the hsp70 family of proteins. Two-dimensional gel elecrophoretic analysis of mutant yeast strains with disrupted *SSA1* and *SSA2* genes (formerly termed YG100 and YG102, respectively (7)) identified the purified proteins as the products of these genes. Deshaies *et al.* performed a genetic analysis of the effect of Ssa1p on translocation of secretory proteins into endoplasmic reticulum and found that this protein also affects translocation *in vivo* (8).

In this manuscript we detail our purification procedure for Ssa1p and Ssa2p (Ssa1p/Ssa2p). We also show that *E. coli dnaK* protein, which is 48% homologous to *Drosophila* hsp70 (9), enhances post-translational translocation of prepro-α-factor.

METHODS

Activity assay

The assay used to purify Ssa1p/Ssa2p has been described (4,5). In summary, translation of prepro-α-factor was directed by SP6 derived mRNA in a wheat germ translation system in the presence of ^{35}[S]-methionine; a small amount of this translation product was then added to yeast microsomes in the presence of cycloheximide, an

ATP regenerating system, and a fraction from the purification to be tested for activity. After incubation the reactions were analyzed by SDS-PAGE as described (3). Translocation of 19 kD prepro-α-factor in this system results in cleavage of the signal sequence (10) and addition of 0,1,2 or 3 core oligosaccharides, yielding products which have M_r of 20 kD, 24 kD, 27 kD and 32 kD, respectively. Therefore, enhanced translocation may be detected by a decrease of the 19 kD primary translation product with a concomitant increase in the 20-32 kD cleaved and glycosylated forms.

The purification was started with a post-ribosomal supernatant (PRS) because it was previously shown to contain the majority of the post-translational translocation enhancing activity (4). All purification steps were done at 4°C.

Purification

Differential centrifugation. A yeast cytoplasmic fraction was prepared as described (termed S100, reference 3) except that 1 mM phenylmethylsulfonyl fluoride was included in the lysis buffer. Thirty liters of cells harvested at an OD_{600} of 2 yielded 64 ml of S100 (39 mg protein/ml). This material was centrifuged at 200,000 x g_{avg} (47,000 rpm in a Beckman Ti50.2 rotor) for 3 hr after reaching speed. 53 ml of PRS (15 mg protein/ml) were removed without disturbing the pellet.

Gel filtration. 52 ml of PRS was sieved through a 1 l Sephadex G-25 medium column equilibrated in 20 mM Hepes-KOH pH 7.4, 2 mM $Mg(OAc)_2$, 2 mM DTT (Buffer A) with 150 mM KOAc and 1 mM MgATP. The excluded protein peak fractions were pooled (PRS-DS, 109 ml, 7.5 mg protein/ml).

Anion exchange chromatography. 98 ml of PRS-DS were loaded onto a 20 ml DE-53 (Whatman) column equilibrated in Buffer A with 150 mM KOAc and 1 mM MgATP. The flow rate was two column volumes per hour throughout. The column was washed with 80 ml of equilibration buffer, then protein was eluted with 80 ml of Buffer A with 300 mM KOAc and 1 mM MgATP and then with 80 ml of Buffer A with 1 M KOAc and 1 mM MgATP. The flow-through protein peak fractions were pooled (DEAE-150, 115 ml, 3.6 mg protein/ml) as were the protein peak fractions that eluted at 300 mM KOAc (DEAE-300, 15 ml, 3.9 mg protein/ml) and at 1 M KOAc (DEAE-1000, 10 ml, 5.7 mg protein/ml).

Gel filtration. 14.4 ml of DEAE-300 were sieved through a 180 ml Sephadex G-25 column equilibrated in Buffer A with 10 mM KOAc. The excluded protein peak fractions were pooled (DEAE-300-DS, 28 ml, 2.1 mg protein/ml).

GTP-agarose chromatography. 27 ml of DEAE-300-DS were loaded onto a 10 ml GTP-agarose (Sigma, GTP attached to beaded agarose via ribose hydroxyls with a 6 carbon spacer, 1.2 umole GTP/ml of gel) column equilibrated in Buffer A with 10 mM KOAc. The flow rate was two column volumes per hour throughout. The column was washed with 40 ml of the same buffer and then protein was eluted with 40 ml of the same buffer containing 10 mM MgGTP. The flow-through protein peak fractions were pooled (GTP-FT, 29.5 ml, 1.4 mg protein/ml). The protein peak fractions that eluted with GTP were also pooled (GTP-GTP, 6.6 ml, 0.4 mg protein/ml).

ATP-agarose chromatography. 28.3 ml of GTP-FT was loaded onto a 5 ml ATP-agarose (Sigma, ATP attached to beaded agarose via N6 amino group with an 8 carbon spacer, 1 umole ATP/ml of gel) column equilibrated in Buffer A with 10 mM KOAc. The flow rate was two column volumes per hour throughout. The column was washed with 20 ml of equilibration buffer, then with 20 ml of Buffer A with 1 M KOAc, then with 20 ml of Buffer A with 100 mM KOAc, and finally with 20 ml of Buffer A with 100 mM KOAc and 10 mM MgATP. The flow-through protein peak fractions were pooled (ATP-FT, 29.0 ml, 0.9 mg protein/ml). The protein peak fractions that eluted with 1 M KOAc were pooled (ATP-KOAc, 2.7 ml, 0.3 mg protein/ml) as were the protein peak fractions that eluted with ATP (ATP-ATP, 8 ml, 0.4 mg protein/ml).

Immunoblot

Anti-Ssa1p/Ssa2p antisera was raised in a rabbit by injection of the 70 kD region of SDS-polyacrylamide gels containing purified Ssa1p/Ssa2p (similar to Figure 1, lane ATP-ATP).

Proteins were electroblotted onto nitrocellulose essentially as described (11). The following manipulations were done at room temperature. The nitrocellulose was blocked with 3% BSA in PBS for 2 hr, washed 3 times for 10 min each in PBS, incubated for 2 hr in a 1:1000 dilution of anti-Ssa1p/Ssa2p antisera in PBS with 0.5% Tween 20 (PBST), washed 6 times for 15 min each with PBST, incubated for 2 hr in a 1:1000 dilution of ^{125}I-Protein A in PBST, washed 6 times for 5 min each in PBST, dried and exposed to Kodak XAR-5 film with an intensifying screen at -80°C.

RESULTS

Figure 1 shows the protein profiles of

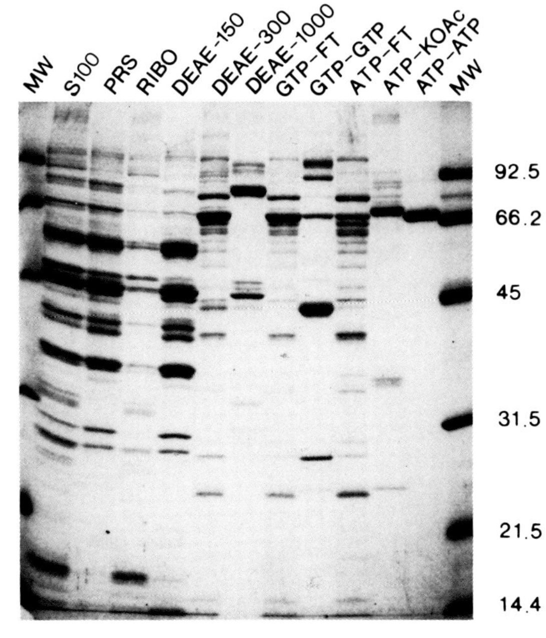

Figure 1. Protein profiles of fractions through the purification of Ssa1p/Ssa2p. SDS-12%PAGE was used and the gel stained with Coomassie blue.

fractions through the purification. The complex pattern of the yeast cytoplasmic fraction is evident (S100, 35 ug protein). The S100 was separated into post-ribosomal supernatant (PRS, 20 ug protein) and ribosomal pellet (RIBO, 20 ug protein) fractions by differential centrifugation. The PRS was then sieved into buffer containing 150 mM KOAc and 1 mM ATP and loaded onto a DEAE anion exchange column. The protein profile of the material that flowed through the column (DEAE-150, 20 ug protein) is significantly different from that of the load. Increasing the KOAc concentration to 300 mM eluted many proteins (DEAE-300, 15 ug protein) among which a 70 kD band is most prominent. A further increase in KOAc concentration to 1M eluted another group of proteins (DEAE-1000, 15 ug protein). The DEAE-300 was then sieved into low-salt buffer without ATP and loaded onto a GTP-agarose column. We used this column to remove several minor contaminants that would have co-purified with the active protein on the next column. Consequently the protein profile of the GTP-agarose column flow through (GTP-FT, 15 ug protein) is rather similar to that of the load. The column was then washed with GTP, releasing a small group of proteins (GTP-GTP, 15 ug protein). Finally, the GTP-FT was loaded onto an ATP-agarose column and most of the proteins flowed through (ATP-FT, 15 ug protein). The column was then washed with 1 M KOAc which eluted several proteins of which a protein of approximately 75 kD was most prominent (ATP-KOAc, 10 ug protein). Elution of the column with ATP released predominantly one 70 kD protein and minor contaminants of higher molecular weight (ATP-ATP, 5 ug protein). As discussed in the introduction and detailed elsewhere (5) this preparation actually consists of both Ssa1p and Ssa2p.

Figure 2 shows a western blot of a gel, identical to that in Figure 1, probed with rabbit antibody raised against Ssa1p/Ssa2p. The Ssa1p/Ssa2p present in the S100 remained predominantly in the PRS, although a significant fraction pelleted with the ribosomes. Co-fractionation with ribosomes is variable and

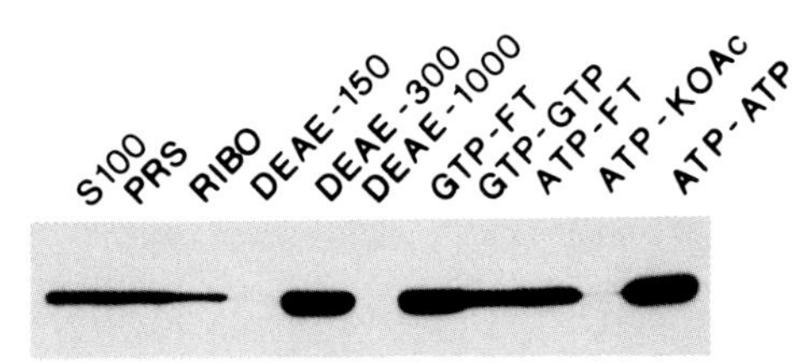

Figure 2. Immunoblot of fractions through the purification with anti-Ssa1p/Ssa2p antisera. Only the 70 kD region of the gel is shown.

unusually high in this particular preparation. The DEAE column gave an essentially quantitative yield of Ssa1p/Ssa2p in the material eluted with 300 mM KOAc. Most of the Ssa1p/Ssa2p loaded onto the GTP-agarose column flowed through, although a significant amount bound and could be eluted with GTP. A large proportion of the Ssa1p/Ssa2p flowed through the ATP-agarose column for reasons that we do not understand but does not seem to be caused by overloading the column. Finally, the Ssa1p/Ssa2p that bound was resistant to elution with 1 M KOAc but was released by elution with ATP. The activity of fractions through the purification corresponds roughly with their content of Ssa1p/Ssa2p (data not shown).

The activity of selected fractions through the purification is shown in Figure 3. The assays shown in lanes 1 and 7 contained no added protein (34% and 31% translocation (Tc), respectively, determined as described in reference 5), that in lane 2 contained 400 ug of S100 (80% Tc), lane 3 had 192 ug PRS protein (84% Tc), lane 4 had 49.6 ug DEAE-300 protein (85% Tc), lane 5 had 22.4 ug of GTP-FT protein (81% Tc), and lane 6 contained 4 ug of ATP-ATP protein (66% Tc). We could not reconstitute translocation as effectively with purified Ssa1p/Ssa2p (ATP-ATP) as with cruder fractions, for example PRS or DEAE-300. This was due to loss of a second factor(s), which is also required for efficient translocation, at the ATP-agarose step (data not shown). We have recently developed an assay for, and are purifying, this second factor. The activity of this component is sensitive to the alkylating agent N-ethylmaleimide, whereas Ssa1p/Ssa2p is resistant (5).

During the purification we noted that when translocation was stimulated with either S100 or

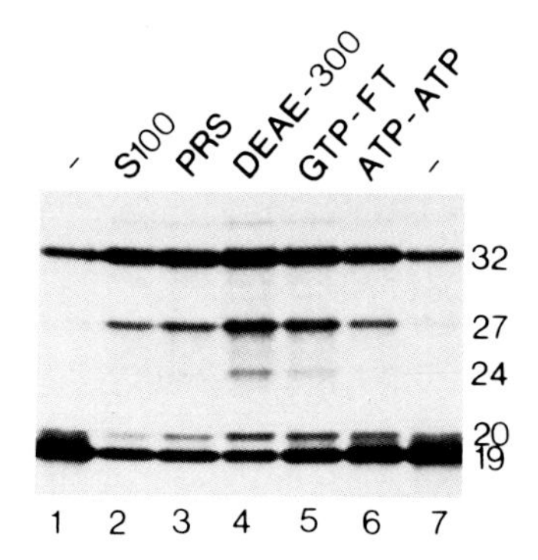

Figure 3. Stimulation of post-translational translocation of prepro-α-factor into yeast microsomes.

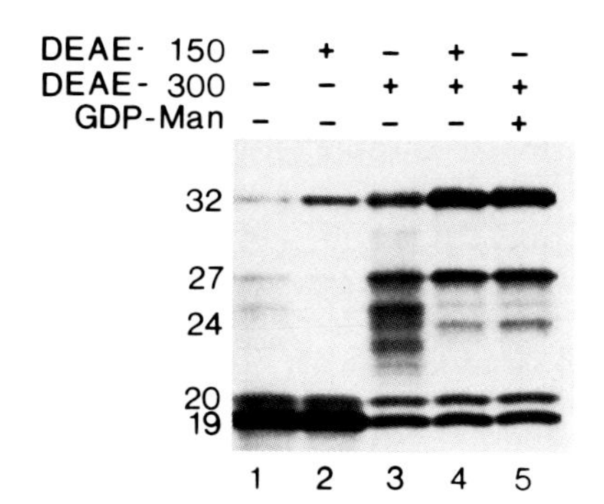

Figure 4. Efficient glycosylation requires a soluble protein or GDP-mannose.

PRS, glycosylation of prepro-α-factor was very efficient. That is, most of the translocated material was evident as the fully glycosylated 32 kD form (similar to Fig. 3, lanes 2 and 3). However, when we used the DEAE-300 fraction glycosylation was very inefficient, as indicated by a heterogeneous population of partially glycosylated forms (Fig. 4, lane 3, 37.6 ug DEAE-300 protein). These glycoproteins were sensitive to endoglycosidase H (data not shown) suggesting that their oligosaccharides extended to at least the $Man_5GlcNAc_2$ form. We noted that the presence of the DEAE-150 fraction during translocation reconstituted efficient glycosylation (Fig. 4, lane 4, 56.0 ug DEAE-150 protein and 37.6 ug DEAE-300 protein). The DEAE-150 fraction alone did not stimulate translocation (Fig. 4, lane 2, 56.0 ug DEAE-150 protein). The activity in the DEAE-150 is, at least in part, proteinaceous because it is sensitive to protease, heat, or N-ethylmaleimide (data not shown). GDP-mannose, which is a precursor of oligosaccharide cores, could substitute for the DEAE-150 in reconstituting efficient glycosylation (Fig. 4, lane 5, 37.6 ug DEAE-300 and 20 uM GDP-mannose). Our data therefore suggest that the active protein in the DEAE-150 is involved in the synthesis of GDP-mannose. We had previously

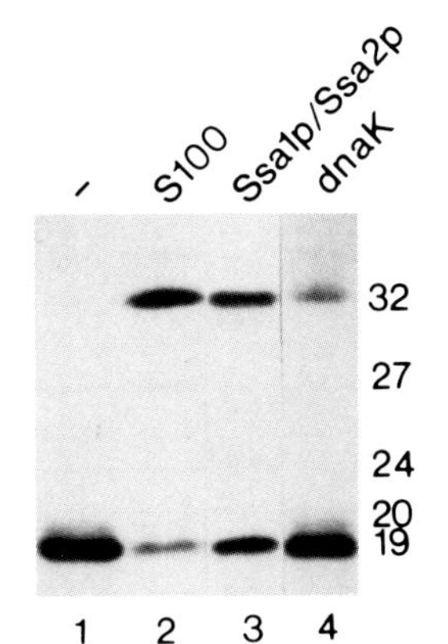

Figure 5. *dnaK* stimulates post-translational translocation of prepro-α-factor.

noticed a similar glycosylation defect when PRS was fractionated on a glycerol gradient and speculated that the defect might be caused by the loss of the *SEC53* (also called *ALG4* (12)) gene product (5), which is involved in efficient glycosylation of secretory proteins (13). Indeed, it has recently been shown that *SEC53* encodes phosphomannomutase, which is required for the biosynthesis of GDP-mannose (14). We have therefore included 20 uM GDP-mannose in our activity assays (Fig. 3) to alleviate the requirement for the active component of DEAE-150, which is probably Sec53p.

In figure 5 we have tested *E. coli dnaK* protein, which is related to Ssa1p and Ssa2p and is essential for replication of bacteriophage lambda (see reference 15 for review), for its ability to enhance post-translational translocation of prepro-α-factor. Lane 1 shows a reaction without added protein, lane 2 a reaction with 6.4 ug of Ssa1p/Ssa2p and lane 3 a reaction with 30.4 ug of purified *dnaK* protein. The *dnaK* protein is capable of stimulating translocation in our system although it is significantly less active than the yeast homolog.

DISCUSSION

We recently developed an *in vitro* assay, based on the post-translational translocation of prepro-α-factor into yeast microsomes, that is enhanced by the presence of soluble proteins (4). We have purified one of the activities by anion exchange and nucleotide affinity chromatography (5). The final preparation consists of two 70 kD ATP-binding proteins with pIs of about 5.5 and 5.4 that have been identified as Ssa1p and Ssa2p, respectively (5). *E. coli dnaK* protein also stimulates post-translational translocation although it is not as active as Ssa1p/Ssa2p.

The mechanism by which these stress proteins facilitate post-translational translocation is not yet understood. They may act to dissociate aggregates (16) of prepro-α-factor that form after synthesis, or "unfold" the precursor prior to, or during, translocation through the membrane, which presumably occurs in an unfolded state (17). The stress proteins may also interact with other components of the translocation apparatus. For example, they may recycle a necessary component from its site of interaction with the microsomal membrane, a function related to the dissociative ability postulated above. Alternatively, stress proteins may associate with a translocation component and maintain it in a native state, possibly by interaction between hydrophobic sites on the two proteins ("protection") and subsequently releasing the component ("deprotection") to perform its function (5). This "protection/deprotection" cycle may require ATP hydrolysis.

ACKNOWLEDGMENTS

We thank Dr. Herbert Weissbach for purified *dnaK* protein.

REFERENCES

1. Hansen W, Garcia PD, Walter P (1986). *In vitro* protein translocation across the yeast endoplasmic reticulum: ATP-dependent post-translational translocation of prepro-α-factor. Cell 45:397-406.
2. Rothblatt JA, Meyer DI (1986). Secretion in

yeast: translocation and glycosylation of prepro-α-factor *in vitro* can occur via an ATP dependent post-translational mechanism. EMBO J 5:1031-1036.

3. Waters MG, Blobel G (1986). Secretory protein translocation in a yeast cell-free system can occur posttranslationally and requires ATP hydrolysis. J Cell Biol 102:1543-1550.
4. Waters MG, Chirico WJ, Blobel G (1986). Protein translocation across the yeast microsomal membrane is stimulated by a soluble factor. J Cell Biol 103:2629-2636.
5. Chirico WJ, Waters MG, Blobel G (1988). 70K heat shock related proteins stimulate protein translocation into microsomes. Nature 332:805-810.
6. Kurtz S, Rossi J, Petko L, Lindquist S (1986). An ancient developmental induction: heat shock proteins induced in sporulation and oogenesis. Science 231:1154-1157.
7. Craig EA, Kramer J, Kosic-Smithers J (1987). *SSC1*, a member of the 70 kDa heat shock protein multigene family of *Saccharomyces cerevisiae*, is essential for growth. Proc Natl Acad Sci USA 84:4156-4160.
8. Deshaies RJ, Koch BD, Werner-Washburne M, Craig EA, Schekman R (1988). A subfamily of stress proteins facilitates translocation of secretory and mitochondrial precursor polypeptides. Nature 332:800-805.
9. Bardwell JCA, Craig EA (1984). Major heat shock gene of *Drosophila* and the *Escherichia coli* heat-inducible *dnaK* gene are homologous. Proc Natl Acad Sci USA 81:848-852.
10. Waters MG, Evans EA, Blobel G (1988). Prepro-α-factor has a cleavable signal sequence. J Biol Chem 263:6209-6214.
11. Towbin H, Staehlin T, Gordon J (1979). Electrophoretic transfer of protein from polyacrylamide gels to nitrocellulose sheets: procedure and some applications. Proc Natl Acad Sci USA 76:4350-4354.
12. Huffaker TC, Robbins PW (1983). Yeast mutants deficient in protein glycosylation.

Proc Natl Acad Sci USA 80:7466-7470.
13. Bernstein M, Hoffmann W, Ammerer G, Schekman R (1985). Characterization of a gene product (Sec53p) required for protein assembly in the yeast endoplasmic reticulum. J Cell Biol. 101:2374-2382.
14. Kepes F, Schekman R. The yeast *SEC53* gene encodes phosphomannomutase. J Biol Chem:in press.
15. Lindquist S (1986). The heat shock response. Ann Rev Biochem 55:1151-1191.
16. Pelham HRB (1986). Speculations on the functions of the major heat shock and glucose regulated proteins. Cell 46:959-961.
17. Eilers M, Schatz G (1986). Binding of a specific ligand inhibits import of a purified precursor protein into mitochondria. Nature 322:228-232.

Stress-Induced Proteins, pages 175–185

THE MOLECULAR ORGANIZATION OF THE SMALL HEAT-SHOCK PROTEINS IN DROSOPHILA

Ch. Haass, P.E. Falkenburg, and P.-M. Kloetzel*

Molekulare Genetik / ZMBH, University of Heidelberg, Im Neuenheimer Feld 282, FRG-6900 Heidelberg

*To whom correspondence should be sent

SUMMARY The molecular organization of the small hsps in Drosophila has been analysed. Our data show that the small hsps synthesized during early pupal stages form globular cytoplasmic 16s particles. Upon heat-shock the 16s particles aggregate into large fast sedimenting complexes. This ability of the small hsps to form 16s particles and large aggregates is independent of the mode of their induction. Our data suggest a common function for the differently induced small hsps which may not be restricted to heat-shock.

INTRODUCTION

The synthesis of the small heat shock proteins of Drosophila melanogaster underlies complex regulatory mechanisms and can be induced by environmental stress, during the development of the fly and by α-ecdysone in Schneider's S3 tissue culture cells (1, 2). The synthesis of the small hsps has been correlated with the gain of increased thermal tolerance. This apparent correlation between the synthesis of the small hsps and the achievement of thermal tolerance seems not to be an unique feature of

Drosophila. A Dictyostelium mutant for example which is deficient for the synthesis of several of the small hsps has lost the ability to survive extreme temperatures (3). Tomato cells also require the synthesis of the small hsps to achieve thermal tolerance (4). But there also exist exceptions for this correlation as in the case of yeast where a deficiency in the synthesis of hsp26 has no obvious effect on the stress resistance of the organism (5). So far little is known about the relationship between the small hsps which are synthesized under the various induction regimes.

Our data show that the small hsps synthesized unter physiological growth condition during normal fly development form soluble globular particles which upon heat shock aggregate into large fast sedimenting complexes. Since the ability of the small hsps to form such complexes is independent of the mode of their induction our data suggest that the small hsps synthesized under the different growth conditions must possess very similar if not identical biological functions.

Material and Methods

Cell culture and growth conditions

The Drosophila melanogaster tissue culture cell line Schneider S3 was used. Cells were grown at 23°C as described before (11). For ecdysone treatment, cells were grown in the presence of 1μM α-ecdysone.

Heat shock of culture cells, labeling, and heat shock recovery

Heat shock of culture cells and labeling conditions were as described previously (10). For heat shock recovery experiments ^{35}s-methionine labeled cells were washed twice in normal growth medium containing an excess of unlabeled L-methionine prior to chase and recovery at 23°C.

Isolation of 16s particles, protein analysis, and electronmicroscopy

Experimental procedures for the isolation and purification of the 16s particles were identical to those described for the isolation of ring-type 19s scRNP particles (11). Proteins were analysed using the Laemmli-system (12) or the protocol of O'Farrell (13). Electron microscopic analysis of the 16s particles was performed as reported before (14).

RESULTS

For analysis of the molecular organization of the developmentally induced small hsps pupal tissues were homogenised and the non polysomal 40000 g supernatant applied to sucrose gradient centrifugation (Fig. 1). The distribution of the small hsps throughout the gradient was an-

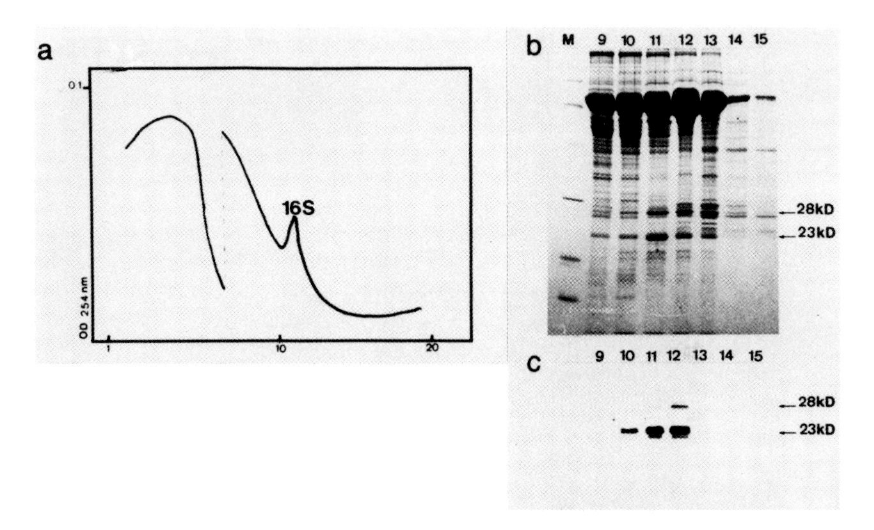

FIGURE 1. Analysis of the 40000g supernatant of homogenates of early pupae by sucrose gradient centrifugation; a) the absorption profile at 254 nm is shown; b) analysis of the proteins contained in the indicated gradient fractions; c) immunoblot using antibody Ec23.

alysed by immunoblotting using the polyclonal antibody Ec23. This antibody identifies hsp23 and to a lesser extent hsp28/27. As shown in Fig. 1c the small hsps which are synthesized during early pupal stages form large complexes sedimenting at ca. 16s in the sucrose gradient. This is in good agreement with previously obtained data which show that α-ecdysone induced small hsps of Schneider's S3 tissue culture cells also possess the ability to form 16s complexes (6). Since the 16s small hsp complexes are stabile in high salt extraction in 0.5 M KAc followed by a second round of high salt sucrose gradient centrifugation was used as an additional purification step. Such purified 16s complexes possess an unchanged protein composition as revealed by 2-dimensional gel electrophoretic separation of the small hsps (Fig. 2b).

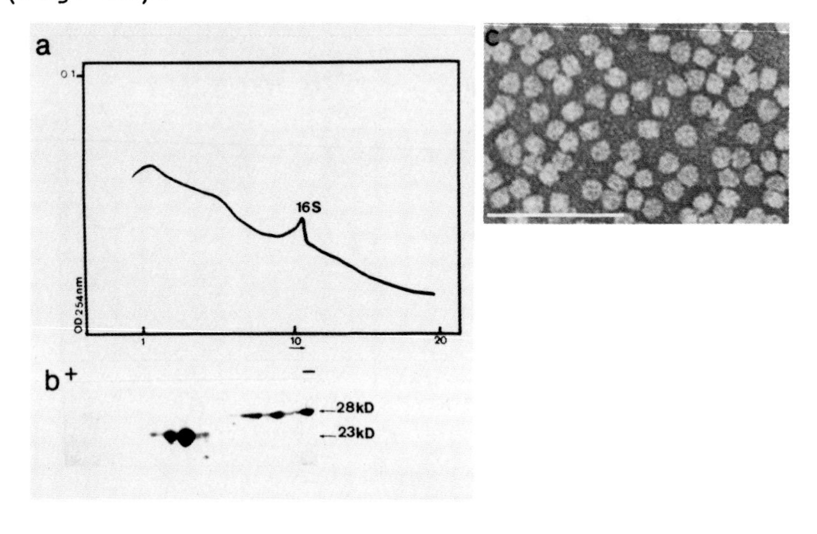

FIGURE 2. Analysis of purified 16s particles:
a) sucrose gradient profile obtained after high salt extraction of the 16s particles and recentrifugation; b) 2-dimensional gel-electrophoretic analysis of the 16s particle proteins; c) electronmicroscopy analysis of the 16s particles. Bar = 1 μm.

Electron microscopic analysis of the pupal 16s particles shows that the small hsps of early pupae form globular particles which possess an average diameter of 12nm.
The ability to form 16s particles is not restricted to α-ecdysone and developmentally induced small hsps but is also a feature of the small hsps which are synthesized after heat-induction (Fig. 3a-c). When the cytoplasmic supernatant of ^{35}S-methionine labeled heat-shocked S-3 tissue culture cells, which have been allowed to recover from heat shock for 6 h at 23°C under chase conditions, is analysed by sucrose gradient centrifugation, the stress

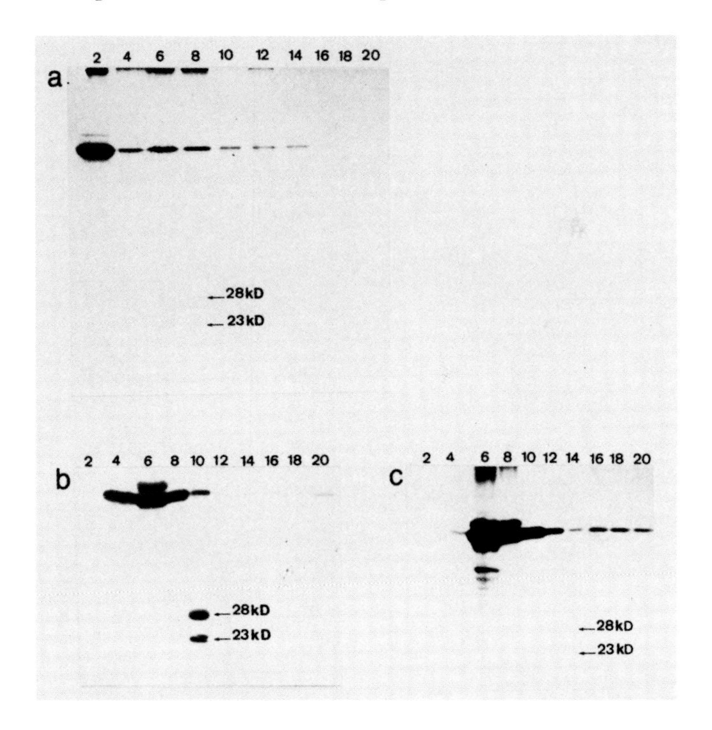

FIGURE 3. Analysis of the ^{35}S-labeled proteins in the 40000g supernatant of S-3 culture cells after:
a) 1h 37°C heat-shock; b) after 6h of recovery at 23 °C; c) after a second 20 min. heat-shock at 37°C.

induced small hsps can exclusively be found sedimenting in the 16s fraction of the gradient. Interestingly when the cytoplasmic supernatant is analysed directly after a 1h heat-shock the labeled small hsps are exclusively recovered from the 40000g pellet and no hsp can be found sedimenting at 16s (Fig. 3a). The same result is obtained when the culture cells are exposed to a second heat-shock following the 6 h recovery period (Fig. 3c).

This ability to shift between different cellular subfraction is not restricted to the heat induced small hsps but can also be observed when early pupae (data not shown) and hormone treated cells are exposed to a short heat-shock at 37°C (Fig. 4a/b). Although α-ecdysone leads to a strong accumulation of

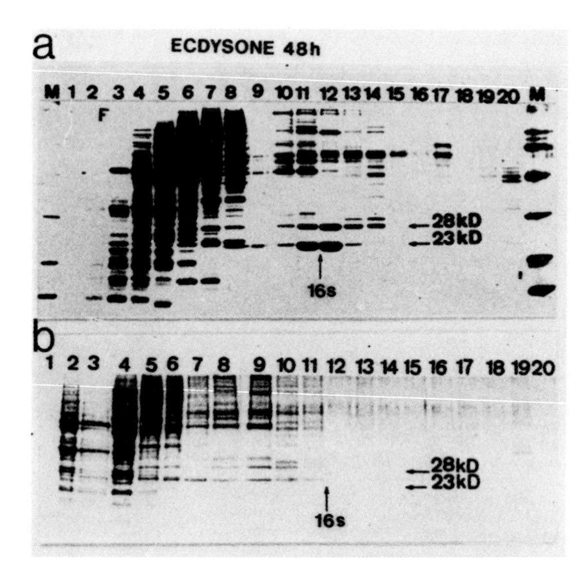

FIGURE 4. Sucrose gradient analysis of the 40000g nonpolysomal supernatant of hormone treated cells before and after heat-shock. Proteins were analysed across the wohle sucrose gradient:
a) Protein content in the gradient fractions before heat-shock; b) protein distribution across the gradient after a 37°C heat-shock.

small hsps in the 16s fraction at normal growth temperature (Fig. 4a) the 40000g supernatant becomes completely devoid of the small hsps after a short exposure of the cells to 37°C. Analysis of the 40000g pellet of heat-shocked ^{35}S-methionine labeled culture cells and of shortly heat-shocked early pupal shows that in both instances the small hsps sediment hetero-

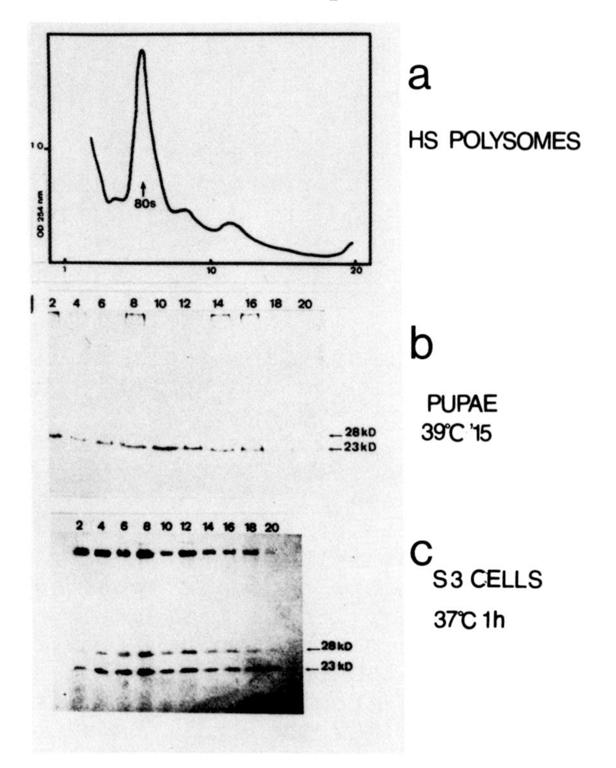

FIGURE 5. Sucrose gradient analysis of polyribosomes extracted in the presence of Triton X-100 from the 40000g pellet of heat-shocked pupae and ^{35}S-labeld culture cells: a) Absorption profile at 254 nm; b) distribution of hsp23 throughout the polysome gradient of heat-shocked pupae; hsp23 was identified by immunoblotting with Ec23; c) distribution of heat-induced ^{35}S-labeled proteins throughout the polysome gradient.

genously as large aggregates with Triton X-100 extractable polyribosomes (Fig. 5a-c). For the interpretation of the latter experiment it is important to note that under these short and severe heat-shock conditions no immunological amounts of stress induced small hsps are synthesized and that the hsps containing large complexes remain stabile even under conditions when polyribosomes are completely dissociated.

Discussion

Our experiments were initiated by the idea that the small hsps independent of their different modes of induction may possess very similar or even identical biological functions.

Previously it was shown that in Drosophila the synthesis of the small hsps underlies several from each other independent regulatory mechanisms and that the synthesis of the small hsps is strongly enhanced during early pupal stages of normal fly development (7).

In this report we demonstrate that concomitant with the increased activation of the small heat-shock genes during normal fly development globular cytoplasmic 16s particles accumulate, which contain the small heat shock proteins as major protein constituents. The formation of globular 16s particles appears to be a general phenomenon of the small hsps since 16s complexes can also be isolated from early embryos or ecdysone treated culture cells. This raises the question how the small hsps which are synthesized under normal physiological conditions relate to the stress induced small hsps. Our experiments show that after 6 hours of recovery from a 1 hour heat shock, the small hsps which had been induced by stress form 16s particles during the recovery period. However, it is interesting to note that directly after heat-shock the 40000g supernatant is completely devoid of small hsps and 16s particles. Right after heat shock the stress induced small hsps as well as the developmentally and α ecdysone induced small hsps can be recovered almost exclusively from the 40000g pellet. A likely interpretation of this observation is that after stress induction the 16s particles and hence the small hsps have the tendency to form large hsp aggregates which in turn results in the cosedimentation of the small hsps with polyribosomal material. However, this does not imply that the small hsps containing aggregates interact directly with the polyribosomes. Such a view would be in line with experimental evidence obtained in tomato cells where after heat shock the small hsps strongly accumulate in the cytoplasm forming large, fast sedimenting heat shock granular (4, 8), while after recovery from heat shock, the small hsps can be isolated from postpolysomal fractions (Nover, pers. commun.).

Once formed, the heat induced 16s particles exhibit the same behaviour as do the developmentally and ecdysone induced 16s complexes. This poses questions as at which molecular level the small hsps unfold their activity and whether the stress induced and developmentally controlled small hsps possess identical targets. Data by Peterson and Mitchell (9) show that one of the prerequisites for the acquisition of thermal tolerance is the ability of a cell to protect and translate 25°C specific mRNA during a stress situation. That is the half life of 25°C mRNA is increased and its translatability during heat shock is improved. These data combined with the observation that the increase in thermal tolerance during Drosophila development correlates well with the induction of small hsp synsthesis, and previous data which show that the small hsps coisolate with poly A^+ RNP (10) suggest that the small hsps may help to provide thermotolerance by interfering with mechanisms which act at the mRNP or translational level.

Since the synthesis of the samll hsps is differentially regulated during fly development, the physiological gain of stress resistance may only be a coincident effect of normal metabolic requirements. In how far, however, the small hsps and the 16s particles are involved in this process remains to be seen.

ACKNOWLEDGEMENT

This work was supported by the Deutsche Forschungsgemeinschaft (SFB 229, C4/Kl).

REFERENCES

1. Zimmerman, J.L. Petri, W., and Meselson, M. (1983). Accumulation of a specific subset of Drosophila melanogaster in RNAs in normal development without heat-shock. Cell 32, 1161-1170.

2. Ireland, R.C. & Berger, E.M. (1982). Synthesis of the low molecular weight heat-shock peptide stimulated by ecdysterone in a cultured Drosophila cell line. Proc.Natl. Acad.Sci.USA 79, 855-859.

3. Loomis, W.F. and Wheeler, S. (1980). Heat-shock response in Dictyostelium. Dev. Biol. 79, 399-408.

4. Nover, L. Scharf, K.D. and Neumann, D. (1983). Formation of cytoplasmic heat-shock granules in tomato cells. Mol.Cell.Biol. 3, 1648-1655.

5. Petko, L. and Lindquist, S.(1986). HSP26 is not required for growth at high temperature, nor for thermotolerance, spore development and germination. Cell 45, 885-894.

6. P.-M. Kloetzel, Ch. Haass, P.E. Falkenburg (1988). The small hsps of Drosophila melanogaster form globular cytoplasmic 16s RNP particles. J.Cell.Biochem. Sup. 12 D, 251.

7. Mason, P.J., Hall, L.M.C. & Gausz, J. (1984). The expression of heat-shock genes during normal development in Drosophila melanogaster. Mol. Gen. Genet. 194, 73-78.

8. Nover, L. & Scharf, K.D. (1984). Synthesis modification and structural binding of heat-shock proteins in tomato cell cultures. Eur. J. Biochem. 139, 303-308.

9. Peterson, N.S. & Mitchell, H.K. (1981). Recovery of protein synthesis after heat-shock: Prior treatment affects the ability of cell to translate mRNA. Proc. Natl. Acad. Sci. USA 78, 1708-1711.

10. Kloetzel, P.M. & Bautz, E.K.F. (1983). Heat-shock proteins are associated with hnRNA in Drosophila melanogaster tissue culture cells. EMBO J. 2, 705-710.

11. Kloetzel, P.M., Falkenburg, P.E., Hössl, P. and Glätzer, K.H. (1987). The 19s ring-type particles of Drosophila: Cytological and biochemical analysis of their intracellular association and distribution. Exp.Cell.Res. 170, 204-213.

12. Laemmli, U.K. (1970). Cleavage of structural proteins during the assembly the head of bacteriophage T4. Nature 227, 680-685.

13. O`Farrell, P.Z., Goodman, H.M., and O`Farrell, P.H. (1977). High resolution two dimensional electrophoresis of basic as well as acidic proteins. Cell, 12, 1133-1139.

14. Schuldt, C. & Kloetzel, P.M. (1985). Analysis of 19s ring-type particles in Drosophila which contain hsp 23 at normal growth temperature. Dev. Biol. 110, 65-74.

Stress-Induced Proteins, pages 187–202

STRUCTURE AND FUNCTION OF MAMMALIAN STRESS PROTEINS[1]

W.J. Welch, L.A. Mizzen, and A.-P. Arrigo

Cold Spring Harbor Laboratory
Cold Spring Harbor, New York 11724

INTRODUCTION

Within the last 2-3 years we have seen a tremendous increase in our understanding of the structure and function of the so-called heat shock or stress proteins. Much of this new information has resulted from the fact that most all of the stress proteins are present in appreciable levels in cells grown under normal conditions. Consequently, many of the stress proteins have begun to be examined by investigators who, at first, were unaware that the protein under study was in fact a member of the stress protein family. Examples of this include the low molecular weight hsp whose synthesis is elevated in cells following exposure to steroids (William McGuire, personal communication) or whose phosphorylation is increased in cells following the addition of growth factors or mitogens (1); the 70 kDa family of stress proteins which has now been shown to be involved in the translocation of proteins from the cytoplasm across the membrane of either the endoplasmic reticulum or mitochondria (2,3); and the 90 kDa hsp which has been shown to be a component of most steroid receptor complexes as well as involved in the transport and/or regulation of various protein kinases (4-6).

In addition to providing new insight on the function of the stress proteins, these studies illustrate the importance of the stress proteins in mediating a number of processes essential to the lifestyle of the normal (unstressed) cell. Whether the elevated synthesis of these proteins after stress is important in restoring these various activities, or whether the proteins have as yet

[1]This work was supported by NIH grant GM33551 to W.J.W.

other defined roles specific to the stressed cell still remains to be established. Work in our own laboratory has centered around defining pertinent morphological and biochemical changes which occur in mammalian cells after stress and identifying, purifying and characterizing the individual stress proteins in mammalian cells. In what follows we discuss some of our current knowledge of the biology of the stressed cell, placing an emphasis on the properties and possible functions of the individual mammalian stress proteins.

RESULTS AND DISCUSSION

The Mammalian Stress Proteins

In defining the stress proteins, we now divide the family into two groups: those referred to as the heat shock proteins and those referred to as the glucose regulated proteins. For example, cells exposed to elevated, nonphysiological temperatures or to heavy metals commonly exhibit an increased synthesis of proteins with apparent masses of 28, 32, 56, 72, 73, 90, and 110 kDa with a concomitant reduction in the synthesis of most other proteins actively being translated prior to the environmental insult. This group of proteins we refer to as the heat shock proteins (Fig. 1). The second family of stress proteins, the glucose regulated proteins, exhibit increased synthesis in cells either deprived of glucose or oxygen or following treatment with agents which perturb calcium homeostasis. These proteins exhibit apparent masses of approximately 75, 80 and 100 kDa (Fig. 1). As can be seen in the 2D gels shown in Figure 1: (i) all of the major stress-induced proteins are relatively acidic polypeptides with their isoelectric points falling between pH 5-7; (ii) almost all of the stress proteins are synthesized at modest or even high levels in cells incubated under normal growth conditions; (iii) synthesis of most (but not all) normal cellular proteins (i.e., those synthesized in the unstressed cell) is decreased in the cells experiencing stress. It should be noted that in cells treated with the amino acid analog of proline, L-azetidine 2-carboxylic acid (Azc), increased synthesis of both the heat shock and glucose regulated proteins is observed (Fig. 1). To our knowledge, this is the only example where both groups of stress proteins are induced simultaneously. Finally, the two groups of proteins are often observed to be regulated in inverse fashion. For example, in cells deprived of glucose or treated with calcium ionophore, there occurs both an increased synthesis of the grps and a concomitant decreased synthesis of some of the hsps (7). Refeeding the cells with glucose or removal of the ionophore results in a reduction of grp synthesis and a corresponding increased production of the hsps (8).

Relationships between the GRP and HSP Families

Biochemical, immunological and cDNA sequence analyses have revealed that members of the two stress protein families are related to one another. Specifically, both the grp 75 and 80 kDa proteins exhibit considerable homology with the hsp 70 proteins (9 and Elizabeth Craig, personal communication), and all of these proteins exhibit binding to ATP *in vitro* (10). Similarly, the grp 100 kDa protein displays high sequence homology with the hsp 90 kDa protein (11). Consequently, because of their similarities, it follows that what we learn about one protein (e.g., grp 80 kDa) may be applicable to its related partner (e.g., hsp 70). As a result, our task in determining the function of the stress proteins has become simplified to some extent. Indeed as will be discussed below, many of the stress proteins which exhibit sequence homology appear to be functioning in a similar fashion but within different compartments in the cell.

Owing to the brevity of this report, I have compiled a table summarizing much of what we know concerning the structure and properties of the individual mammalian stress proteins. This data, summarized in Table 1 and discussed in more detail below, consists of a composite of studies performed in our and many other laboratories and represents a brief update of the biology of the individual stress proteins of mammalian cells.

The Heat Shock Proteins

<u>28 kDa</u>. Does not label efficiently with [^{35}S]methionine and therefore is most commonly detected using a mixture of [^{3}H]amino acids. In cells grown at 37°C, steady-state synthesis of 28 kDa is low and therefore is

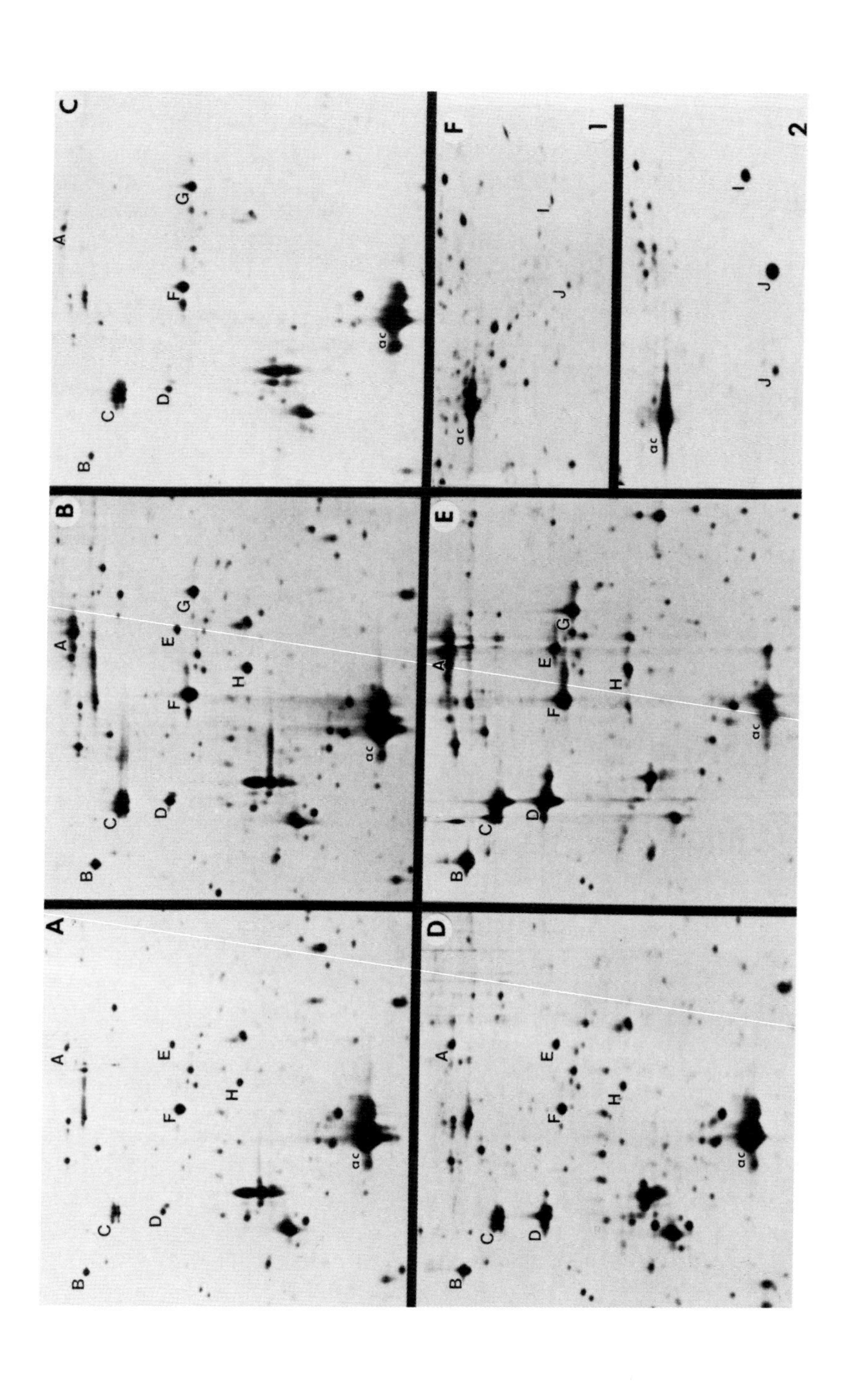

generally difficult to detect. In contrast, the protein is easily visualized in the 37°C cell following labeling with $[^{32}P]H_3PO_4$ (1,12). The protein consists of at least four isoforms, three of which contain phosphate (1,12). More interesting is the fact that 28 kDa phosphorylation, but not its synthesis, increases significantly in cells exposed to a variety of different mitogens or tumor promoters (1). Hence, it will be interesting to see if the mammalian 28 kDa, like the Drosophila small molecular weight heat shock proteins, exhibits differential expression and/or changes in phosphorylation during early development and/or differentiation (13,14). The protein has been purified from Hela cells and exhibits a rather high native molecular mass ranging from 200-400,000 daltons in 37°C cells on up to $>2x10^6$ daltons in cells after heat shock (15). Like the low mw hsps from other organisms, 28 kDa shows sequence homology to the α-crystallin proteins present in the lens (16,17). The α-crystallins similarly form higher ordered structures, and we suspect, therefore, that those domains in common between the the proteins are responsible for the self-assembly properties of the two proteins.

Using an antibody raised against the purified 28 kDa protein we have observed, via indirect immunofluorescence, a perinuclear distribution of 28 kDa in cells maintained at 37°C. Using double-label staining with a fluorescent lectin which recognizes Golgi glycoproteins, we have shown that much of this perinuclear staining of 28 kDa is coincident with the Golgi complex (18). Our preliminary conclusions

---

FIGURE 1. Stress proteins of mammalian cells. Rat embryo fibroblasts (REF-52) were labeled with $[^{35}S]$methionine at: A, 37°C; B, after a 43°C/90 min heat shock; C, after exposure to 120μM sodium arsenite; D, after exposure to 10μM calcium ionophore A23187; E, after exposure to the proline analogue L-azetidine 2 carboxylic acid (5mM). After labeling, the cells were harvested and the labeled proteins analyzed by two-dimensional gel electrophoresis (pH 4-8, acid end to the left). Shown in panels A-E are only those regions analyzing the higher molecular weight stress proteins. In panel F is a portion of the gels showing the lower molecular weight heat shock proteins (1, normal; 2, arsenite-treated cells). Stress proteins indicated are: A, 110 kDa; B, 100 kDa, C, 90 kDa; D, 80 kDa; E, 75 kDa; F, 73 kDa; G, 72 kDa; H, 58 kDa; I, 32 kDa; J, 28 kDa. Actin is indicated, ac.

are that 28 kDa is not an integral Golgi protein but rather appears to be present around the Golgi stacks. Following heat shock treatment, much of 28 kDa localizes within the nucleus and to a lesser extent is dispersed throughout the cytoplasm. Such dispersal of the protein may be correlated with our previous observations demonstrating that much of the Golgi complex breaks up after heat shock treatment (19). As the cells recover from the shock, there is an exit of the protein from the nucleus and a corresponding gradual return of the protein into the perinuclear region. Our biochemical studies examining the solubility of 28 kDa are consistent with our immunological localization studies. While the protein is easily solubilized by detergent treatment of the 37°C cells, much of 28 kDa is found in the insoluble nuclear/cytoskeletal fraction after heat shock. With time of recovery from heat shock, the protein is observed to gradually return to the detergent soluble fraction. Interestingly, in cells first made thermotolerant and then exposed to a second heat shock treatment, significantly less of 28 kDa partitions into the detergent insoluble fraction (18).

32 kDa. The induction of this protein appears somewhat limited to agents which affect sulfhydryl groups. For example, various heavy metals and thiol reactive agents such as iodoacetamide result in a significant induction of 32 kDa as well as the other classical heat shock proteins (20,21). In a limited number of cells heat shock treatment does result in a slight induction of the 32 kDa protein (Mizzen and Welch, unpublished observation). Recent data has also shown induction of the 32 kDa protein in cells exposed to ultraviolet light or the oxidizing agent H_2O_2 (22). Little else is known about the properties or intracellular locale of this somewhat unusual stress protein.

58 kDa. Another relatively recent addition to the stress protein family, 58 kDa, was first reported as a heat shock protein in Tetrahymena (23). The protein is present within the mitochondria and is not apparently exposed to the cytoplasmic side of the mitochondrial membrane as ascertained by proteolytic protection experiments. We have recently identified and purified a homolog of 58 kDa from mammalian cells (24). 58 kDa is made as a precursor protein, and upon insertion into the mitochondria, the protein is processed into its final mature form. The protein exists as an oligomer exhibiting a sedimentation value of ≅20-25S. In this respect it is interesting that 58

kDa appears homologous to the bacterial hsp GroEL (25), a protein which similarly forms large aggregates. Finally, recent data has shown that GroEL also exhibits similarity with an abundant chloroplast protein involved in the binding and perhaps assembly of the photosynthetic CO_2-fixation protein, Rubisco (26). Hence it will be interesting to see if the hsp 58 kDa protein, like the Rubisco-binding protein, plays a role in mediating the assembly of macromolecular complexes within the mitochondrion.

72 kDa and 73 kDa. Of all of the stress proteins, we know the most concerning the so-called "70 kDa" stress protein members. As was mentioned earlier, there appear to be two major members of this family: the rather abundant 73 kDa protein whose synthesis is readily apparent in all cells grown under normal conditions. Consequently, 73 kDa is referred to as the "cognate" or "constitutive" 70 kDa stress protein. In contrast, the second member, what we refer to as 72 kDa, is not obvious in most cells grown at 37°C. Instead synthesis of 72 kDa, in general, occurs only after induction of the stress response and in most cases represents the major translational activity of the stressed cell. An exception to this rule occurs in primate cells. So far in all primate cells examined (over 15 human and 3 monkey cell lines), synthesis of the 72 kDa protein is observed in the cells grown under normal conditions (7 and unpublished observation). Following heat shock treatment, increased synthesis of 72 kDa and to a lesser extent 73 kDa is observed. A number of interesting observations concerning this constitutive expression of 72 kDa in primate cells has emerged: (i) its expression appears cell-cycle regulated with synthesis being observed at the Gi/S boundary (27); (ii) 72 kDa expression, in general, appears to increase in primate cells following transformation (Welch, unpublished observation); and (iii) the expression of 72 kDa increases in human cells (but not rodent cells) in response to transfection (or infection) with either of two cooperating oncogenes, *E1A* or *myc* (28-30).

Immunological, biochemical and DNA sequence analysis has demonstrated that while the 72 kDa and 73 kDa proteins are highly related, they are in fact distinct gene products (reviewed in 31 and 32). In addition, both proteins show very similar biochemical properties including their stoichiometric copurification during either gel filtration or ion exchange chromatography (33). Finally both proteins exhibit a high affinity for various nucleotides, the highest being for ATP. This property of the proteins has facilitated their relatively simple and rapid purification (10).

Recent observations from a number of laboratories have begun to point toward a biochemical role for the hsp 70 proteins. First, two reports have shown that these proteins, *in vitro*, facilitate the uncoating or release of clathrin triskelions from clathrin coated vesicles (34-35). Such activity appears ATP dependent. Second, data from both Blobel's laboratory and from Craig and Schekman's laboratory have implicated a role for the 70 kDa heat shock proteins in the translocation of secretory and mitochondrial precursor proteins from their site of synthesis in the cytoplasm across the endoplasmic reticulum and mitochondria, respectively (2,3). Again it seems that these translocation events require ATP and perhaps ATP hydrolysis. Finally, the grp 80 kDa protein, homologous to the hsp 70 proteins, is identical to the BiP protein present within the endoplasmic reticulum and which appears to interact with abnormal or incompletely assembled proteins in an ATP-dependent manner (9,46-48). Thus a picture is emerging (discussed further below) in which the hsp 70 proteins appear to interact, in a transient manner, with a number of other proteins within either the cytosolic or nuclear compartments. Moreover, these interactions may involve an ATP-dependent folding/unfolding of the target protein. Having described some of the properties of the 70 kDa hsps in the normal cell, the question then arises as to their behavior in the cell experiencing stress.

Biochemistry of the hsp 70 Proteins in the Cell Experiencing Stress

After stress, synthesis of 72 kDa represents the major translation product of the cells. 72 kDa consists of multiple related isoforms, the exact number depending upon the cell type, the agent used to induce the response and finally the severity of the stress treatment (1,7). Biochemical fractionation and immunological studies have shown that the majority of 72 kDa localizes within the nucleus shortly after its synthesis (36,37). Here the protein appears present within the nucleoplasm, the nuclear matrix, and large amounts accumulating within the nucleoli (37). This nucleolar deposition of the protein always appears correlated with marked alterations in the integrity of the nucleoli. Such alterations in the nucleoli appear to coincide with a diminishment of nucleolar function, most notably the shutdown of rRNA synthesis and ribosomal assembly (38,39). Using immunoelectron microscopy, we have shown that most of the nucleolar distributed 72 kDa is present within the so-called granular region--that area of the nucleolus involved in the assembly of small ribonucleoproteins and pre-ribosomes (40). During recovery from the heat shock treatment, the nucleoli slowly regain their normal morphology (and function), and 72 kDa is observed to exit the organelle (40,41).

During the later periods of recovery from physiological stress, the majority of 72 kDa begins to accumulate within the cytoplasm. Three distinct cytoplasmic locales are observed: a strong perinuclear distribution; a portion present within unusually large and phase dense structures; and finally a portion present directly underneath the plasma membrane, again in association with phase dense material. Using electron microscopy and double-label indirect immunofluorescence analysis, we have found that much of the cytoplasmic 72 kDa is in very close association with ribosomes and polysomes (40). This colocalization of 72 kDa with the polysomes, we think, is integral to the functioning of the translational machinery during and after recovery from heat shock treatment.

90 kDa. The mammalian 90 kDa stress protein, homologous to the Drosophila 83 kDa stress protein, is a very abundant protein in cells grown under normal conditions and whose synthesis increases approximately 3-5 fold after induction of stress. 90 kDa is heavily phosphorylated with there being at least 12 isoforms, half of which appear to contain phosphate (7). The protein has been purified from the cytosolic fraction of cells, exists as a dimer (and perhaps higher molecular mass forms) and copurifies with a small amount of the 100 kDa stress protein (33). 90 kDa has also been shown to interact with a number of other interesting intracellular proteins. For example, the protein displays a transient interaction with a number of tyrosine kinases (see review by 6). Specifically, newly synthesized kinases, such as $pp60^{src}$, associate with 90 kDa and another cellular protein of 50 kDa. While present in this complex, the *src* protein shows neither tyrosine kinase activity nor phosphorylated tyrosine residues. As the complex containing $pp60^{src}$ reaches its final destination at the plasma membrane, the complex dissociates with *src*, being deposited at the inner side of the membrane and the protein now exhibiting both tyrosine kinase activity and

autophosphorylated tyrosine. In a somewhat analogous situation, 90 kDa is found associated with the 8S form of various steroid hormone receptor complexes (4,5). Upon binding of the steroid, there is a conversion of the receptor to the biologically active 4S form with an accompanying loss of the 90 kDa stress protein. Finally, very recent data implicate a role for 90 kDa in the regulation of yet another protein kinase. Specifically, 90 kDa copurifies with the heme-regulated protein kinase which phosphorylates the alpha subunit of the eukaryotic initiation factor 2 (eif-2α) (42). Eif-2α is involved in the regulation of translational initiation and exhibits an increased level of phosphorylation in some cells following heat shock treatment (43). Increased phosphorylation of eif-2α appears to result in a decrease of ribosomal initiation complexes and the subsequent inhibition of protein synthesis. These studies, in sum then, point toward a role of 90 kDa in regulating the activity of other important macromolecules in the cell.

110 kDa. Of all of the major members of the heat shock protein family, we know the least regarding the properties of this protein. 110 kDa is a constitutively expressed protein whose synthesis increases approximately five-fold after stress. The protein is present within the nucleolus in what appears to be the fibrillar region, i.e., that site within the nucleolus involved in rRNA transcription (44).

The Glucose-regulated Proteins

75 kDa. Very recent studies in our laboratory have defined a new stress protein whose synthesis appears affected by many, but not all, of the treatments which increase the synthesis of the two major glucose-regulated proteins (80 kDa and 100 kDa, discussed below). This protein, we refer to as 75 kDa, is a mitochondrial protein. 75 kDa appears to yet another member of the hsp 70 family. The protein binds to ATP and exhibits immunological cross-reactivity with some antibodies against the hsp 70 family. Like the mitochondrial hsp 58 kDa protein, 75 kDa is made as a precursor and is processed to its mature form presumably as it crosses into the organelle (24). Current studies are in progress to determine whether 75 kDa serves a role in the assembly/disassembly of mitochondrial proteins.

80 kDa. 80 kDa is a relatively abundant protein in cells grown under normal conditions. The protein contains

phosphate and/or is ADP-ribosylated and has been shown to be present within the endoplasmic reticulum (7,9,45). As was discussed earlier, 80 kDa is homologous to hsp 70 and similarly binds ATP (10). Studies by Pelham have shown that grp 80 kDa is in fact identical to a protein, BiP, first observed as an ER protein which transiently interacts with maturing immunoglobulin heavy chains (9,46). It also binds to a number of mutant or abnormal proteins which traffic through the endoplasmic reticulum (47,48). Similar to that described above for the 70 kDa proteins, it appears that grp 80 (or BiP) functions in an ATP-dependent manner to facilitate the correct folding and/or assembly of proteins which pass through the ER.

100 kDa. The second major member of the glucose-regulated protein family, 100 kDa, contains both phosphate and carbohydrate and has been localized to both the ER, Golgi, and plasma membrane (7). As was mentioned above, grp 100 kDa exhibits considerable sequence homology with hsp 90, a cytosolic protein (11). Although little is known concerning 100 kDa function, we suspect that owing to its sequence homology with 90 kDa, we might expect 100 kDa to similarly exhibit transient interactions with other macromolecules, such as protein kinases, for example, but within its own compartment, the ER and Golgi.

A Family of ATP-binding Proteins and Their Possible Role in the Stressed Cell

Having described some of the properties of the stress proteins, especially as they pertain to their role in cells grown under normal conditions, the question arises as to their function in the cell after heat shock or other types of physiologic stress. Our best clues to answering this question follow from two observations: (i) that there exists a related family of stress proteins, all of which are capable of binding ATP but which are compartmentalized within different locales in the cell and (ii) the phenomenon of thermotolerance; the ability of cells to withstand and survive what would otherwise be a lethal heat shock challenge providing that the cells have been given a prior mild stress exposure. With respect to the former, it is clear now that there exists at least four major members of the 70 kDa stress protein family: the constitutive 73 kDa, the highly stress-induced 72 kDa, the glucose-regulated 80 kDa (BiP), and finally a new member, 75 kDa. Each of these family members appear to reside in different compartments: 72 and 73 kDa in the cytoplasm and nucleus, 80 kDa within the endoplasmic reticulum and 75 kDa within the mitochondria. Despite their different locales, each of these proteins appear to be performing similar tasks: the folding/unfolding of target proteins, presumably catalyzed by the energy of ATP hydrolysis. For example, the 72/73 kDa proteins have been shown to be involved in the ATP-dependent disruption of clathrin coated vesicles (34,35) and recently the ATP-dependent translocation of newly synthesized secretory and mitochondrial proteins from their site of synthesis in the cytoplasm across their appropriate target membrane, the endoplasmic reticulum and mitochondrion, respectively (2,3). In the case of grp 80 kDa, the protein appears to regulate the proper assembly of immunoglobulin heavy and light chain as well as perhaps involved in the binding of abnormal or incompletely assembled protein within the endoplasmic reticulum, again in an ATP-dependent manner (9,46-48). Finally, although just recently identified, we suspect that the mitochondrial form of hsp 70, the 75 kDa protein, may, like grp 80 kDa, be involved in regulating the assembly of multi-subunit structures within the mitochondria utilizing the energy of ATP hydrolysis. In all cases then, these related proteins appear to be involved in transient protein-protein interactions, most of which involve folding/unfolding events and all of which are most likely ATP dependent.

One suspects then that after heat shock or treatment of cells with other stress agents, most of which are protein denaturants, the cell increases the production of a group of proteins whose primary role is concerned with the refolding and/or removal of denatured proteins whose concentration is increasing throughout various cellular compartments in the stressed cell. Credence for this idea is the observation that the 72/73 kDa stress proteins following heat shock interact with denatured and/or aggregated preribosomes within the nucleolus and apparently facilitate the restoration of proper nucleolar function (i.e., ribosome assembly) (40,41). We might expect, therefore, that each of the 70 kDa protein family members, within their respective compartments, will similarly function to facilitate the recovery of essential metabolic functions: protein translocation across the membrane, protein trafficking through the endoplasmic reticulum, and protein assembly within the mitochondria, activities all of which might be expected to be perturbed in the stressed cell.

TABLE 1. PROPERTIES OF MAMMALIAN STRESS PROTEINS

Stress Protein ($x10^{-3}$ dalton)	Modification	Location 37	Location 42	Remarks
HSP				
28	phosphorylation	cytoplasm near Golgi	nuclear cytoplasm	increased phosphorylation in response to mitogens/tumor promoters; homology to lens alpha crystallin; native mass $\geq$400,000 daltons, forms aggregates
58	phosphorylation?	mitochondria	mitochondria	synthesized as a precursor; homology to E. coli GroEL and to Rubisco-binding protein
72	methylation	(nuclear cytoplasm)	cytoplasm nuclear nucleolar	most induced; low levels in 37°C cells; affinity for nucleotides, RNA, fatty acids; cell-cycle regulated, role in DNA synthesis?; ↑ expression in cells transfected with "cooperating" oncogenes (E1A, SV-40 T, c-*myc*); association with p53, E1A, polyoma m-T
73	methylation	cytoplasm	cytoplasm nuclear nucleolar	abundant constitutive protein related to 72; affinity for nucleotides, RNA, and fatty acids; portion present with microtubules; homology with clathrin-coated vesicle uncoating ATPase
90	phosphorylation methylation	cytoplasm	cytoplasm (nuclear?)	abundant cellular phosphoprotein with multiple isoelectric forms; increased phosphorylation in response to ds DNA; transient association with tyrosine kinases; present in steroid hormone receptor complex; synthesis repressed following glucose/calcium deprivation
110	phosphorylation?	nucleolar	nucleolar cytoplasm	
GRP				
75	--------	mitochondria	mitochondria	synthesized as a precursor; homology to hsp 72/73 and grp 80
80	phosphorylation ADP-ribosylation	endoplasmic reticulum	endoplasmic reticulum	increased synthesis following glucose/calcium deprivation; homology to immunoglobulin heavy chain binding protein; binds ATP; related to HSP 70
100	phosphorylation glycosylation	Golgi/ER plasma membrane	Golgi/ER plasma membrane nuclear	increased synthesis during glucose/calcium deprivation; portion associated with 90kDa stress protein; related to HSP 90

If in fact these activities of the hsp 70 family members are correct, one might envision then how the cell acquires transient resistance to heat. In all organisms and cells it has been shown that exposure of cells to a mild sublethal heat shock treatment and subsequent recovery at the normal temperature result in the cells acquiring resistance to a second and what would otherwise be a lethal heat shock event. We suspect that the basis of acquired thermotolerance is simply the ability of the cells to function normally under conditions which are toxic to the cell. Indeed it has been shown already that functions such as RNA processing, translation and ribosomal assembly are severely inhibited in nontolerant cells after stress but which are protected (and/or repaired faster) in cells first made thermotolerant and then exposed to a second stress event (18,49-52). Presumably the protection and/or maintenance of these (and others to be determined) activities may, in part, be mediated by the various members of the hsp 70 proteins distributed within different compartments throughout the cell. With many of these activities now defined, along with the availability of both the purified hsp 70 family members and their corresponding antibodies, we are now in the position to begin testing these proposed roles of the hsp 70 family of proteins.

ACKNOWLEDGMENTS

We thank Jim Garrels and the Quest Laboratory for 2D gel analysis, P. Renna for photographic work and M. Szadkowski for preparation of the manuscript.

REFERENCES

1. Welch WJ (1985). J Biol Chem 260:3058.
2. Deshaies RJ, Koch BD, Werner-Washburne M, Craig EA, Schekman R (1988). Nature 332:800.
3. Chirico WJ, Waters MG, Blobel G (1988). Nature 332:805.
4. Catelli MG, Binart N, Jung-Testas I, Renoir JM, Baulieu EE, Feramisco JR, Welch WJ (1985). EMBO J 4:3131.
5. Sanchez ER, Toft DO, Schlesinger MJ, Pratt WB (1985). J Biol Chem 260:12398.

6. Brugge JS (1985). In Vogt PD (ed): "Current Topics in Microbiology and Immunology," Germany: Springer Press, vol 122.
7. Welch WJ, Garrels J, Thomas GP, Lin JJC, Feramisco JR (1983). J Biol Chem 258:7102.
8. Subjeck J, Shyy J (1986). Am J Physiol 250:C1.
9. Munro S, Pelham HR (1986). Cell 46:291.
10. Welch WJ, Feramisco JR (1985). Mol Cell Biol 5:1229.
11. Surgan DR, Tsai MJ, O'Malley BW (1986). Biochemistry 25:6252.
12. Kim YJ, Shuman T, Sette M, Przybla A (1984). Mol Cell Biol 4:468.
13. Sirotkin K, Davidson N (1982). Dev Biol 89:196.
14. Arrigo A-P (1987). Dev Biol 122:39.
15. Arrigo A-P, Welch WJ (1987). J. Biol Chem 262:15359.
16. Ingolia TD, Craig EA (1982). Proc Natl Acad Sci USA 79:2360.
17. Hickey ES, Brandom E, Potter R, Stein G, Stein J, Weber LA (1986). Nuc Acids Res 14:4127.
18. Arrigo A-P, Suhan JP, Welch WJ (1988). Manuscript submitted.
19. Welch WJ, Suhan JP (1985). J Cell Biol 101:1198.
20. Levinson W, Oppermann H, Jackson J (1980). Biochim Biophys Acta 606:170.
21. Caltabiano M, Koestler TP, Poste G, Greig RG (1986). J Biol Chem 261:13381.
22. Keyse SM, Tyrrell RM (1987). J Biol Chem 262:14821.
23. McMullin TW, Hallberg RL (1987). Mol Cell Biol 7:4414.
24. Mizzen LA, Garrels JG, Welch WJ (1988). Manuscript submitted.
25. McMullin TW, Hallberg RL (1988). Mol Cell Biol 8:371.
26. Hemmingsen SM, Woolford C, van der Vies SM, Tilly K, Dennis DT, Georgopoulos CP, Hendrix RW, Ellis RJ (1988). Nature 333:330.
27. Milarski KL, Morimoto RI (1986). Proc Natl Acad Sci USA 83:9517.
28. Nevins JR (1982). Cell 29:913.
29. Kingston RE, Baldwin AS, Sharp PA (1984). Nature 312:280.
30. Wu BJ, Hurst AC, Jones NC, Morimoto RI (1986). Mol Cell Biol 6:2994.
31. Craig EA (1985). CRC Crit Rev Biochem 18:239.
32. Lindquist S (1986). Ann Rev Biochem 55:1151.
33. Welch WJ, Feramisco JR (1982). J Biol Chem 257:14949.
34. Chappell TG, Welch WJ, Schlossman DM, Palter KB, Schlesinger MJ, Rothman JE (1986). Cell 45:3.

35. Ungewickel E (1985). EMBO J 4:3385.
36. Velazquez JM, Lindquist S (1984). Cell 36:655.
37. Welch WJ, Feramisco JR (1984). J Biol Chem 259:4501.
38. Amalric F, Simard R, Zalta JP (1969). Exp Cell Res 55:370.
39. Rubin GM, Hogness DS (1975). Cell 6:207.
40. Welch WJ, Suhan JP (1986). J Cell Biol 103:2035.
41. Pelham HR (1984). EMBO J 3:3095.
42. Rose D, Welch WJ, Kramer G, Hardesty B. (1988). Submitted for publication.
43. Duncan R, Hershey JWB (1984). J Biol Chem 262:14538.
44. Subjeck J, Shyy T, Shen J, Johnson RJ (1983). J Cell Biol 97:1389.
45. Carlsson L, Lazarides E (1983). Proc Natl Acad Sci USA 80:4664.
46. Haas IG, Wabl M (1983). Nature 306:387.
47. Gethig MJ, McCammon K, Sambrook J (1986). Cell 46:939.
48. Kassenbrock CK, Garcia PD, Walter P, Kelley RB (1988). Nature 333:90.
49. Yost HJ, Lindquist S (1986). Cell 45:185.
50. Petersen NS, Mitchell HK (1981). Proc Natl Acad Sci USA 78:1708.
51. Mizzen LA, Welch WJ (1988). J Cell Biol 106:1105.
52. Welch WJ, Mizzen LA (1988). J Cell Biol 106:1117.

Stress-Induced Proteins, pages 203–219

STEROID HORMONE RECEPTORS AND HEAT SHOCK PROTEIN M_r 90,000 (HSP 90): A FUNCTIONAL INTERACTION ?

Etienne-Emile Baulieu and Maria-Grazia Catelli

INSERM U 33 and Faculte de Medecine Paris-Sud, Lab. Hormones, 94275 Bicêtre Cedex.

ABSTRACT In the native, heterooligomeric, non active 8S-receptor, hsp 90 caps the receptor DNA-binding site. Hormone (agonist) "transforms" the 8S-form, releasing hsp 90 and active (4S-) receptor which can bind to DNA and trigger the hormonal response. An antihormone (antagonist) as RU 486 stabilizes the 8S-form, and thus decreases availability of the DNA-binding site of the receptor. Hsp 90 may be prototype of "antireceptor" and more generally of transcriptional regulators which do not bind to DNA but interact with the DNA-binding site of regulatory proteins.

Key words: Hsp 90, steroid receptor, antihormone, RU 486, glucocorticosteroid, progesterone, estrogen.

INTRODUCTION

Steroid hormones are transcriptional regulators (1-3). They act within the target cell nucleus via receptors, that are specific proteins to which the hormones bind with high affinity (K_D of the order of $10^{9 \pm 1}$ M) (4-6). Five distinct receptors (R) correspond to the five categories of steroid hormones: estrogen (E), progesterone (P), androgen (A), glucocorticosteroid (G) and mineralocorticosteroid (M) ligands. These receptors have been cloned, their cDNA sequenced, and overall homologies between them have been revealed by deduction of their primary structures (7, 8). Like the thyroid hormone receptor, which is c-erb A itself, and the receptors for calcitriol, retinoic acid and probably other yet unknown ligands, they belong to the erb-A super family of ligand responsive transcription factors (review in 9).

a) Functional domains of steroid receptor. Schematically, four domains may be described in all steroid receptors (Fig.1). First, there is a putative DNA-binding region, ~ 70 aa long, rich in positively charged amino acids (Lys, Arg), whose cystein pattern suggests two Zn-stabilized finger structures of the 2C-2C type (10,11). From available information (a limited number of animal species have been studied so far, and the detailed sequence of the androgen receptor is not published), homology in this DNA-binding region, which determines the specificity of interaction with hormone regulated elements (HRE) of target genes is ~ 100% for receptors of a given hormone through different mammalian and avian species, and between 50 and 90% in paired comparisons between different steroid hormone receptors. The estrogen receptor is ~ 50% homologous to other receptors in this region, while PR, GR, MR and AR constitute a subfamily with more than 80% homology. This may explain why different receptors can bind to the same HRE of some regulated genes: e.g., PR and GR bind to the same HRE of the lysozyme gene (12); however detailed studies show that the binding of two receptors to the same HRE is not identical (13), and this may be responsible, in part, for the different kinetics and efficiency of the response of the same gene to two different hormones. Thus, at the level of the DNA-binding site of receptors, high homology and receptor-specific differences are probably both involved in the specificity of the hormonal response.

Second, there is a steroid binding region, situated in the C-terminal portion of the receptors, and which include ~ 250 aa with a high proportion of hydrophobic residues. Its entire sequence appears to be involved in hormone binding. Amino acid homology in this region varies from ~ 30% for ER vs GR to > 50% for PR or MR vs GR. High aa homology between, PR, GR and MR correlates with spillover of binding specificity for steroid ligands. In the steroid binding domain, there is a ~ 20 aa zone, ~ 200 aa away from the C-terminal, which is the most conserved sequence in the hormone binding region (between aa 577 and 596 in human GR). Thirdly, between the DNA and the steroid binding sites, the "hinge" region (~ 30 aa) is not significantly conserved, and probably has no role in the information transfer between the hormone site and the DNA-binding region; however, it may play a mandatory role

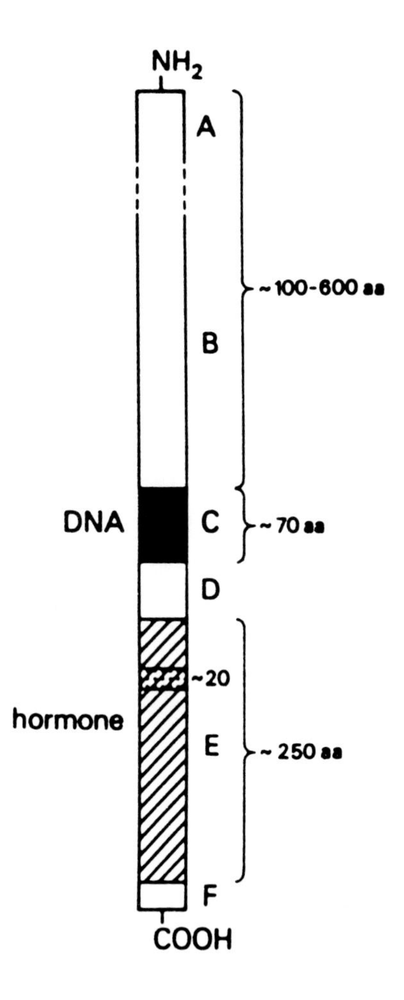

FIGURE 1. Consensus structure of DNA-binding ligand-responsive transcription factor. The super family of these proteins include the five classical steroid hormone receptor, the calcitriol, the thyroid hormone and the retinoic acid receptors. DNA and hormone stand for putative DNA and ligand binding domains. The ~ 20 amino acid region (~ 20) is discussed in the text. A, B, C, D, E, F is the monenclature of receptor regions, used by P. Chambon (8).

in gene expression, possibly implicating clusters of basic amino acids (14). Fourthly, the N-terminal portion of the receptors seems important in determining tissue and target gene specific activation, as demonstrated for the chicken progesterone receptor (15). The length of this region, the least conserved of receptors, is variable (100-600 aa): and accounts for the different molecular weights, from ER: M_r ~ 75 k to MR: M_r ~ 120 k.

b) Non DNA binding, hsp 90 containing, 8S-form of steroid receptors. When target cell extracts are analyzed, steroid receptors detected by radioactive hormonal ligand binding, are found in two forms (4-6). One is obtained in low salt cytosol of tissues or cells, in the absence of the corresponding hormone; it is large, with a sedimentation coefficient of 8-9S ("8S-R") and an apparent MW of ~ 300K, and it does not bind to DNA. The other form is smaller, usually with a sedimentation coefficient of ~ 4S ("4S-R"). It binds to DNA, and is actually extracted by salt-containing buffer from nuclei of target cells which have been exposed to the corresponding steroid hormone. Alternatively it can be obtained in absence of hormone after making the cytosol > 0.3 M KCl, or after salt treatment of the purified 8S-R. Thus it appears that the 8S-R is a supramolecular structure, dissociated by an increase in ionic strength.

Purification of both 8S and 4S forms of the chick oviduct PR by hormonal affinity chromatography permitted the obtention of antibodies. One monoclonal antibody (BF4), raised after 8S-PR injection (16), did not recognize the 4S-PR, neither in Western blot experiments where the receptor is denaturated, nor in density gradients in which it is not denatured. The BF4 monoclonal antibody reacted with a 90K protein component of the 8S-PR, and it was found that the same 90K protein was also included in the 8S form of ER, GR, AR (17) and MR (18) in the chick, and this has been confirmed in all species tested thus far (19-24). The 90K protein present in the 8S-R was identified as the heat shock protein M_r 90,000 (hsp 90) by molecular cloning and immunocharacterization (25-28).

Molybdate and other oxyanions (vanadate and tungstate) (29-31) stabilize the 8S hetero-oligomeric receptors, and cross-linking experiments with bisimidates, in particular dimethylpimelimidate, have established that the interaction between the receptor and hsp 90 detected in cytosol is selective (32). In addition, cross-linking of the 8S PR confirmed results already observed with molybdate-stabilized GR (29), which suggested that the interaction of the receptor with the 90K protein stabilizes the former, i.e., it protects the hormone binding site against thermal inactivation (32). Studies aimed at defining protein interaction and stoichiometry of receptor and hsp 90 in the 8S form, have indicated that hsp 90 is always in a dimeric form, and that it is associated with one molecule of receptor in the case of PR (21) and GR (33), and with two molecules of receptor in the case of ER (22). The 8S complex appears to be held together by relatively weak electrostatic forces: its disruption occurs spontaneously by simply raising the temperature, or can be induced by breaking the electrostatic interactions of the charged regions of the subunits involved, either by KCl or by heparin. Transformation by neither KCl nor heparin depend on the presence of ligand bound to the receptor, and the properties of the receptor molecule produced by treatment of ligand-free receptor with high ionic strengh or with heparin were identical with those of the activated hormone-receptor complex, demonstrating that receptor activation can be obtained experimentally in the absence of hormone (34,35). The stability of the interaction of different steroid receptors with hsp 90 is not the same: for instance, in the presence of the same molybdate ion concentration, ER remains more sensitive to the dissociating effect of increased concentrations of KCl than PR and GR (36). This dissociation is reversible (37,38), but not if the Zn has been removed from the receptor (e.g., by 1,10-phenanthrolin), suggesting that the Zn-finger structure should be maintained for insuring the reconstitution of the 8S form of the receptor as well as the DNA-binding property of the 4S-R (38).

c) A model for receptor-hsp 90 interaction. Cloning and sequencing of chick hsp 90 cDNA (25,39) indicated a 728 aa protein with a MW of ~ 84,000 and a calculated pI of 5.1. Two different isoforms of hsp 90 encoded by different genes, have been found in several species (40-46). In the mouse, where two hsp 90 proteins have been identified (44), both appear to participate in the 8S-GR (46).

All hsps 90 in mammals (43,45), chicken (39), drosophila (48), yeast (49) and trypanosoma (50), include two charged regions A and B. In the chick, they extend between residues 221 and 290, and 530 and 581, respectively (39). While overall aa conservation between eucaryotic hsp 90 is > 65%, the A region of these proteins is the most variable, but the distribution of charged amino acids is remarkably conserved throughout species (Fig.2). It is interesting that the A region is almost absent in the homologous protein of E. Coli (51). A region in chicken hsp 90 (70 aa) has 53 charged aa, of which 33 are acidic and 20 are basic; this excess of negative charges may account for the ionic interaction with the positively charged DNA-binding region of the receptor. Thus the A region was chosen for tridimensionnal modelling presented in Fig.3. Using a new method (52), the predicted structure of hsp 90 A region is an α-helix (aa 233-247), followed, after a proline- containing loop of 13 aa, by a second α-helix (aa 261-287). The polyacidic helices are probably stabilized by the presence of positively charged amino acids, as in troponin C (53). Superimposition of the glutamic and aspartic carboxyl group with phosphates of the backbone of the B-DNA helix occurs in 15 out of 30 negatively charged residues of this region. Therefore, the A region of hsp 90 may be viewed as a DNA-like structure (39), potentially able to cap the DNA-binding site of steroid receptors.

To test this hypothesis an antibody was obtained against a synthetic peptide derived from the A region of hsp 90 (Courtesy of Dr Rolf Geiger, Hoescht, Frankfurt). While this antibody bound to free hsp 90, it did not bind to the hsp 90 A region included in the receptor 8S-form, a finding consistent with the proposed model of interaction between the receptor and hsp 90 (54) (Fig.4).

The availability of mutated receptor cDNAs of the human GR (55) and ER (56) suggested further studies. The receptors were expressed in monkey cells devoid of receptor (i.e., Cos), and their interaction with endogenous hsp 90 was analysed. Preliminary results with ER cDNA lacking the putative DNA binding site confirm the importance of this region in the formation of the 8S-ER structure (B. Chambraud, personal communication). With GR (57), results indicated that a deletion of the DNA-binding region makes the 8S-form unstable; in addition the steroid binding region also seemed to be involved in the interaction with hsp 90. GR mutants, truncated just upstream of the most conserved 20 aa in the steroid binding domain or largely deleted in the same domain, did not form an 8S-heterooligomer.

```
chick        (221)  VEKERDKEVS DDEA-EE-KE EEKEEKEE-- K-TEDKPEIE DVGSDEEEEK KDGD-KKKK- KKIKELYIDE EELNKTK
mouse        (205)  LEKXPEKEIS DDEA-----E EEKGEKEEXD KEXEEKPKIE DVGSD EEDD SGKD-KKKKT KKIKEKYIDQ EELNKTK
human        (217)  LEKEPEKEIS DDEA-----E EEKGEKEEED KDDEEKPKIE DVGSD-EEDD SGKD-KKKKT KKIKEKYIDQ EELNKTK
drosophila   (210)  VEKEPEKEVS DDEADDE-KK EGDEKK-E-- MET-DEPKIE DVGEDEDADK KDKDAKKKKT --IKEKYTED EELNKTK
yeast        (209)  VTKEVEKEVP IPE--EEKKD EEK-KDEEKK DEDDKKPKLE EV--DEEEE- -----KKPKT KKVKEEVQEI EELNKTK
trypanosoma  (208)  VEKATEKEV- TDE-DE--DE AAATKNEE-- GE-EPKVE-E -VKDDAFE-- --G-EKKKKT KKVKEVTQEF VVQNKHK
E. coli      (210)  IEKREEKDGE TVISWEK--- ---------- ---------- ---------- ---------- ---------- --INKAQ
```

FIGURE 2. Negative charge conservation in the A region of seven hsps 90. The amino acid sequences are aligned to give maximum identities. References are cited in the text. Numbers in parentheses refer to the first aa of the A region. Negatively charged amino acids are represented by boldfaced letters. X's in the mouse sequence indicate undetermined aa.

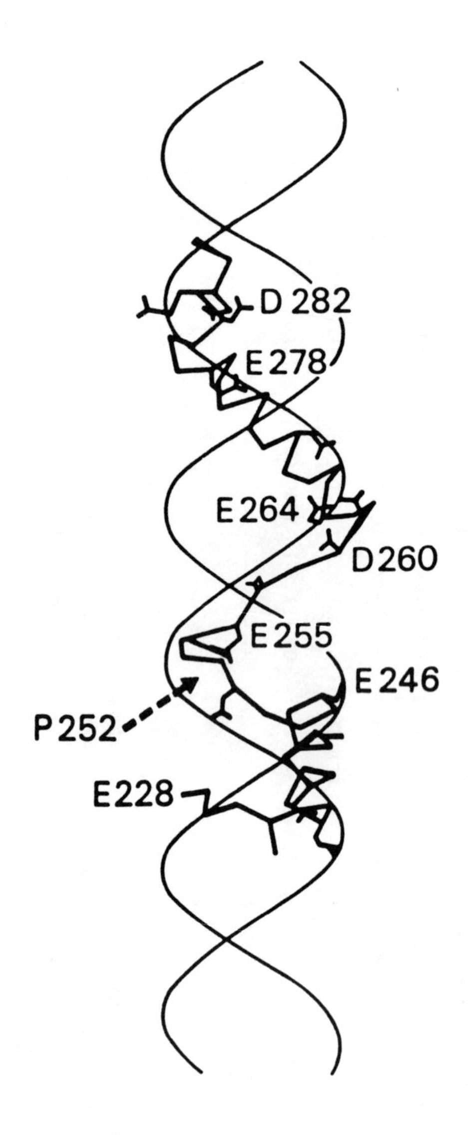

FIGURE 3. Modelling of zone A aa 228-290 of chick hsp 90 (39).

Further experiments are needed to fully delineate the mode of interaction of hsp 90 with the DNA-binding region of the receptor, as well as for defining the possible interaction of hsp 90 with other regions of the receptor. This other type of interaction may facilitate the ionic binding of hsp 90 region A with the receptor DNA-binding site.

In conclusion, the bulk of evidence obtained by studying the physicochemical properties of the 8S-receptor components, the results of the immunological studies and the experiments with mutated receptor support strongly the model involving ionic interaction between the positively charged Zn-finger structure of receptors and the DNA-like A region of hsp 90.

d) Functional aspects. We have already discussed (58,59) that the 8S receptor form exists in living cells, and is not an artefact of homogeneization favoring the interaction of receptor with the more abundant hsp 90 (0.1-1% of total soluble protein in most cells). This has been recently confirmed by pulse-chase labeling experiments (60,61). In absence of hormone, PR and ER completely and GR in part appear to be nuclear proteins, as indicated by immunocytochemical data (review in 62). This observation is consistent with the localization of some hsp 90 in nuclei of several cell types (63) (it is usually believed that hsp 90 is exclusively a cytoplasmic protein (64); only after heat shock of chick embryo fibroblasts was hsp 90 previously observed in the nucleus (65)).

In absence of hormone, hsp 90 binding to the receptor may have some function of protection (see above) and transport to the nucleus of the receptor synthetized in the cytoplasm. In addition, it prevents the receptor to interact with DNA. We then believe that an essential step in hormone action is that steroid binding transforms the physicochemical characteristics of the receptor, relieves the inhibitory effect of hsp 90 on DNA binding, and thus triggers the hormonal response. The double observation that some C-terminal truncated steroid receptors are active constitutively (i.e. in absence of hormone they cannot bind anyway) (55) and that they do not form an 8S structure (57), is consistent with the proposed role for hsp 90 to obliterate the DNA-binding site of receptors in absence of hormone. It has been also been observed that the binding of the antihormonal steroid RU 486 stabilizes the interaction between GR and PR with hsp 90, and such an effect may be involved in the anticorticosteroid and antiprogesterone activities of the compound (66,67). Contrary to what occurs in presence of a glucocorticosteroid agonist, GR in living cells does not protect HRE of the TAT gene from chemical methylation when it binds RU 486 (68). These observations again are supporting the proposed mechanism of action of steroid hormones and antihormones presented in Fig.4. However, up to now, antiestrogens have not displayed any stabilizing effect on 8S-ER, and this may be related to the different stability of the 8S-form of ER vs GR and PR, as reported above.

Since hsp 90, in the preceding exemples, appears to regulate negatively the receptor function, it may be envisaged that its increased concentration relative to that of the receptor participates in a deficit of hormone action in some cases of hormone resistance, despite sufficient level of receptor. This seems to be possible in the mouse mammary gland during lactation, where estrogen in sensitivity correlates with impeded in vitro activation of ER and with the maximum level of hsp 90 (70).

In summary, we suggest that: 1. In the native, heterooligomeric, non active 8S-receptor, hsp 90 caps the receptor DNA-binding site. 2. Hormone (agonist) "transforms" the 8S-form, releasing hsp 90 and active (4S-)receptor which can bind to DNA and trigger the hormonal response. 3. An antihormone (antagonist) as RU 486 stabilizes the 8S-form, and thus decreases availability of the DNA-binding site of the receptor. 4. hsp 90 may be prototype of "antireceptor" and more generally of transcriptional regulators which do not bind to DNA but interact with the DNA-binding site of regulatory proteins.

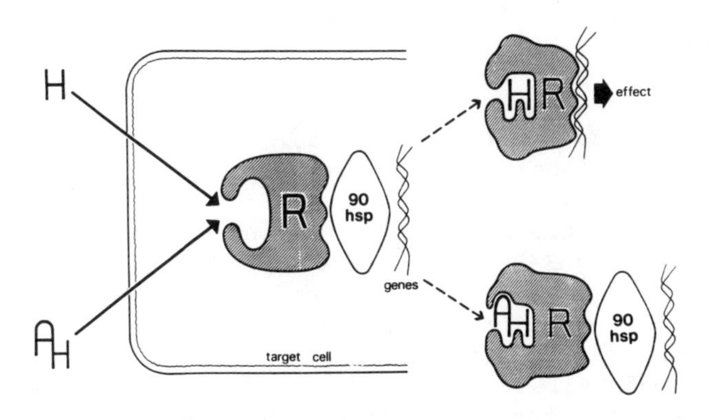

FIGURE 4. Schematic representation of the receptor system and its transformation upon binding of agonist (hormone H) or antagonist (antihormone AH). In absence of hormones, the receptor is under heterooligomeric 8S form, containing R and hsp 90. Hsp 90 caps the receptor in binding to the DNA-binding site (ionic bonds), as discussed in the text. The stoichiometry of the complex in not represented. The probable interaction of hsp 90 with R, elsewhere than in the DNA-binding region, is not indicated. No interaction of the 8S-receptor with DNA takes place. In case of hormone binding, hsp 90 is released; the 4S-receptor interacts with DNA and triggers the response at the DNA level. The changes of conformation when R binds to H or AH are not indicated.

REFERENCES

1. Yamamoto KR (1985). Steroid receptor regulated transcription of specific genes and gene networks. Annu Rev Genet 19:209-252.
2. Bishop JM (1985). Viral oncogenes Cell 42:23-30.
3. Ringold GM (1985). Steroid hormone regulation of gene expression. Annu Rev Pharmacol Toxicol 25:529-566.
4. Jensen EV, Suzuki T, Kawashima T, Stumpf WE, Jungblut PW, DeSombre ER (1968). A two-step mechanism for the interaction of oestradiol with the rat uterus. Proc Natl Acad Sci USA 59:632-638.
5. Gorski J, Toft DO, Shyamala G, Smith D, Notides A (1968). Hormones receptors: studies on the

interactions of estrogen with the uterus. Rec Progr Hormone Res 24:45-80.

6. Baulieu EE, Alberga A, Jung I, Lebeau MC, Mercier-Bodard C, Milgrom E, Raynaud JP, Raynaud-Jammet C, Rochefort H, Truong H, Robel P (1971). Metabolism and protein binding of sex steroids in target organs: an approach to the mechanism of hormone action. Rec Progr Hormone Res 27:351-419.
7. Hollenberg SM, Weinberger C, Ong ES, Cerreli G, Oro A, Lebo R, Thompson EB, Rosenfeld MG, Evans RM (1985). Primary structure and expression of a functional human glucocorticoid receptor cDNA. Nature 318:635-641.
8. Green S, Chambon P (1986). A superfamily of potentially oncogenic hormone receptors. Nature 324:615-617.
9. Evans, RM (1988). The steroid and thyroid hormone receptor superfamily. Science 240:889-895.
10. Miller J, McLachlan AB, Klug A (1985) Repetitive zinc-binding domains in the protein transcription factor IIIA from Xenopus oocytes. EMBO J 4:1609-1614.
11. Berg JM (1988). Proposed structure for the zinc-binding domains from transcription factor IIIA and related proteins. Proc Natl Acad Sci USA 85:99-102.
12. Cato ACB, Miksicek R, Schütz G, Arnemann J, Beato M (1986). The hormone regulatory element of mouse mammary tumour virus mediates progesterone induction. EMBO J 5:2237-2240.
13. Von der Ahe D, Renoir JM, Buchou T, Baulieu EE, Beato M (1986). Receptors for glucocorticosteroid and progesterone recognize distinct features of a DNA regulatory element. Proc Natl Acad Sci USA 83:2817-2821.
14 Adler S, Waterman ML, Xi He, Rosenfeld MG (1988). Steroid receptor-mediated inhibition of rat prolactin gene expression does not require the receptor DNA-binding domain. Cell 52:685-695.
15. Tora L, Gronemeyer H, Turcotte R, Gaub MP, Chambon P (1988). The N-terminal region of the chicken progesterone receptor specifies target gene activation. Nature 333:185-188.
16. Radanyi C, Joab I, Renoir JM, Richard-Foy H, Baulieu EE (1983). Monoclonal antibody to chicken oviduct progesterone receptor. Proc Natl Acad Sci USA 80:2854-2858.

17. Joab I, Radanyi C, Renoir JM, Buchou T, Catelli MG, Binart N, Mester J, Baulieu EE (1984). Immunological evidence for a common non hormone-binding component in "non-transformed" chick oviduct receptors of four steroid hormones. Nature 308:850-853.
18. Oblin ME, Couette B, Radanyi C, Lombes M, Baulieu EE (1988). Mineralocorticoid receptor of the chicken intestine : oligomeric structure and transformation (submitted) J Biol Chem.
19. Riehl RM, Sullivan WP, Vroman BT, Bauer VJ, Pearson GR, Toft DO (1985). Immunological evidence that the nonhormone binding component of avian steroid receptors exists in a wide range of tissues and species. Biochemistry 24:6586-6591.
20. Housley PR, Sanchez ER, Westphal HM, Beato M, Pratt WB (1985). the molybdate-stabilized L-cell glucocorticoid receptor isolated by affinity chromatography or with a monoclonal antibody is associated with a 90-92-kDa nonsteroid-binding phosphoprotein. J Biol Chem 260:13810-13817.
21. Renoir JM, Buchou T, Mester J, Radanyi C, Baulieu EE (1984). Oligomeric structure of the molybdate-stabilized, non-transformed "8S" progesterone receptor from chicken oviduct cytosol. Biochemistry 23:6016-6023.
22 Redeuilh G, Moncharmont B, Secco C, Baulieu EE (1987). Subunit composition of the molybdate-stabilized "8-9" non-transformed estradiol receptor purified from calf uterus. J Biol Chem 262:6969-6975.
23. Mendel DB, Bodwell JE, Gametchu B, Harrison RW, Munck A (1985). Molybdate-stabilized nonactivated glucocorticoid-receptor complexes contain a 90-kDa non-steroid-binding phosphoprotein that is lost on activation. J Biol Chem 261:3758-3763.
24. Ghering U, Arndt H (1985). Heteromeric nature of glucocorticoid receptors. Febs Letters 179:138-142.
24b. Renoir JM, Buchou T, Baulieu EE (1986). Involvement of a non-hormone binding 90kDa protein in the non-transformed 8S form of the rabbit uterus progesterone receptor. Biochemistry 25:6405-6413.
25. Catelli MG, Binart N, Feramisco JR, Helfman D (1985). Cloning of the chick hsp 90 cDNA in expression vector. Nucl Acid Res 13:6035-6047.
26. Catelli MG, Binart N, Jung-Testas I, Renoir JM, Baulieu EE, Feramisco JR, Welch WJ (1985). The common 90-kd protein component of non-transformed "8S"

steroid receptors is a heat-shock protein. EMBO J 4:3131-3135.
27. Sanchez ER, Toft DO, Schlesinger MJ, Pratt WB (1985). Evidence that the 90-kDa phosphoprotein associated with the untransformed L-cell glucocorticoid receptor is a murine heat-shock protein. J Biol Chem 260:12398-12401.
28. Schuh S, Yonemoto W, Brugge J, Bauer VJ, Riehl RM, Sullivan WP, Toft DO (1985). A 90,000-dalton binding protein common to both steroid receptors and the Rous sarcoma virus transforming protein, $pp60^{v-src}$. J Biol Chem 260:14292-14296.
29. Nielson CJ, Sando JJ, Vogel WM, Pratt WB (1977). Glucocorticoid receptor inactivation under cell-free conditions. J Biol Chem 252:7568-7578.
30. Toft D, Nishigori H (1979) Stabilization of the avian progesterone receptor by inhibitors. J Ster Biochem 11:413-416.
31. Wolfson A, Mester J, Yang CR, Baulieu EE (1980). "Non-activated" form of the progesterone receptor from chick oviduct: characterization. Biochem Biophys Res Commun 127:71-79, 1982.
32. Aranyi P, Radanyi C, Renoir M, Devin J, Baulieu EE (1988). Covalent stabilization of the nontransfomed chick oviduct cytosol progesteorne receptor by chemical cross-linking. Biochemistry 27:1330-1336.
33. Denis M, Wikström AC, Gustafsson JA (1987). The molybdate-stabilized non-activated glucocorticoid receptor contains a dimer of M_r 90,000 non-hormone-binding protein. J Biol Chem 262:11803-11806.
34. Milgrom E, Atger M, Baulieu EE (1973). Acidophilic activation of steroid hormone receptors. Biochemistry 12:5198-5205.
35. Yang CR, Mester J, Wolfson A, Renoir JM, Baulieu EE (1982). Activation of the chick oviduct progesterone receptor by heparin in the presence or absence of hormone. Biochem J 208:399-406.
36. Redeuilh G, Secco C, Baulieu EE, Richard-Foy H (1981). Calf uterine estradiol receptor: effects of molybdate on salt induced transformation process and characterization of a non-transformed receptor state. J Biol Chem 256:11496-11502.
37. Rochefort H, Baulieu EE (1971). Effect of KCl, $CaCl_2$, temperature and oestradiol on the uterine cytosol receptor of oestradiol. Biochimie 53:893-907.

38. Sabbah M, Redeuilh G, Secco C, Baulieu EE (1987). The binding activity of estrogen receptor to DNA and heat shock protein (M_r 90,000) is dependent on receptor-bound metal. J Biol Chem 262:8631-8635.
39. Binart N, Chambrau B, Dumas B, Rowlands DA, Bigogne C, Levin JM, Garnier J, Baulieu EE, Catelli MG (1988). The cDNA-derived amino acid sequence of chick heat shock protein M_r 90,000 (hsp 90) reveals a "DNA like" structure: a potentiel site of interaction with steroid receptors. (submitted) J Biol Chem.
40. Morange M, Diu A, Bensaude O, Babinet C (1984). Altered expression of heat shock proteins in embryonal carcinoma and mouse early embryonic cells. Mol Cell Biol 4:730-735.
41. Hickey E, Brandon SE, Sadis S, Smale G, Weber LA (1986). Molecular cloning of sequences encoding the human heat-shock proteins and their expression during hyperthermia. Gene 43:147-154.
42. Hickey E, Smale G, Lloyd D, Weber LA (1988). J Cell Biochem, supplement 12D, abstract 278 p 312.
43. Rebbe NF, Ware J, Bertina RM, Modrich P, Stafford DW (1987). Nucleotide sequence of a cDNA for a member of the human 90-kDa heat-shock protein family. Gene 53:235-245.
44. Ullrich SJ, Robinson EA, Law LW, Willingham M, Appella E (1986). A mouse tumor-specific transplantation antigen is a heat shock-related protein. Proc Natl Acad Sci USA 83:3121-3125.
45. Moore SK, Kozak C, Robinson EA, Ulrich SJ, Appella E (1987). Cloning and nucleotide sequence of the murine hsp84 cDNA and chromosome assignment of related sequences. Gene 56:29-40.
46. Lindquist S (1986). The heat-shock response. Ann Rev Biochem 55:1151-1191.
47. Mendel DB, Orti E (1988). Isoform composition and stoichiometry of the ~ 90-kDa heat shock protein associated with glucorticoid receptors. J Biol Chem 263:6695-6702.
48. Blackman RK, Meselson M (1986). J Mol Biol 188:499-515.
49. Farrelly FW, Finkelstein DB (1984). Complete sequence of the heat shock-inducible HSP90 gene of saccharomyces cerevisiae. J Biol Chem 259:5745-5751.
50. Dragon EA, Sias SR, Kato EA, Gabe JD (1987). teh genome of trypanosoma cruzi contains a constitutively expressed, tandemly arranged multicopy gene

homologous to a major heat shock protein. Mol Cell Biol 7:1271-1275.

51. Bardwell JCA, Graig EA (1987). Eukariotic M_r 83,000 heat shock protein has a homologue in Escherichia coli. Proc Natl Acad Sci 84:5177-5181.

52. Biou V, Gibrat JF, Levin JM, Robson B, Garnier J (1988). Secondary structure production: combination of thre methods Protein Engeneiring 2, in press.

53. Sundaralingam M, Sekharudu YC, Yathindra N, Ravichandran V (1987). Ion pairs in alpha-helices. Proteins Str Fun Gen 2:64-71.

54. Catelli MG, Radanyi C, Renoir JM, Binart N, Baulieu EE (1988). Definition of domain of hsp 90 interaction with steroid receptors. UCLA Symposia on Stress-Induced Proteins, April 10-16, Keystone, in press.

55. Giguere V, Hollenberg SM, Rosenfeld MG, Evans RM (1986). Functional domains of the huamn glucocorticoid receptor. Cell 46:645-652.

56. Kumar V, Green S, Saub A, Chambon P (1986). Localization of the oestradiol-binding and putative DNA-binding domains of the human oestrogen receptor. EMBO J, 5: 2231-2236.

57. Pratt WB, Jolly DJ, Pratt DV, Hollenberg SM, Giguere V, Cadepond F, Schweizer-Groyer G, Catelli MG, Evans RM, Baulieu EE (1988). A region in the steroid binding domain determines formation of the non-DNA-binding, 9S glucocorticoid receptor complex. J Biol Chem 263:267-273.

58. Baulieu EE, Binart N, Buchou T, Catelli MG, Garcia T, Gasc JM, Groyer A, Joab I, Moncharmont B, Radanyi C, Renoir JM, Tuohimaa P, Mester J (1983). In Eriksson H, Gustafsson JA (eds): "Steroid Hormone Receptors: Structure and Function. Elsevier, Amsterdam: p 45-72.

59. Baulieu EE (1987). Steroid hormone antagonists at the receptor level. A role for the heat-shock protein MW 90,000 (hsp 90). J Cell Biochem 35:161-174.

60. Mendel DB, Bodwell JE, Munck A (1987). Activation of cytosolic glucocorticoid-receptor complexes in intact WEHI-7 cells does not dephosphorylate the steroid-binding protein. J Biol Chem 262:5644-5648.

61. Howard KJ, Distelhorst CW (1988). Evidence for intracellular association of the glucocorticoid receptor with the 90-kDa heat shock protein. J Biol Chem 263:3474-3481.

62. Gasc JM, Baulieu EE (1987). Intracellular localization of steroid hormone receptors: immunocytochemical analysis. In Moudgil VK (ed): "Proceeding of the Meadow Brook conference on steroid receptors in health and disease." New York, (in press).
63. Gasc JM, Renoir JM, Radanyi C, Joab I, Tuohimaa P, Baulieu EE (1984). Progesterone receptor in the chick oviduct: an immunohistochemical study with antibodies to distinct receptor components. J Cell Biol 99:1193-1201.
64. Schlesinger MJ, Ashburner M, Tissieres A (1982). "Heat Shock from bacteria to man" Cold Spring Harbor Laboratory NY.
65. Collier NC, Schlesinger MJ (1986). The dynamic state of heat shock proteins in chicken embryo fibroblasts. J Cell Biol 103:1495-1507.
66. Groyer A, Schweizer-Groyer G, Cadepond F, Mariller M, Baulieu EE (1987). Antiglucocorticoid effects suggest why steroid hormone is required for receptors to bind DNA in vivo but not in vitro. Nature 328:624-626.
67. Renoir JM, Radanyi C, Devin J, Baulieu EE (1988). The antiprogesterone RU486 stabilizes the heterooligomeric, non-DNA binding, 8S-form of the rabbit uterus cytosol progesterone receptor. Biochem J. (submitted).
68. Becker PB, Gloss B, Schmid W, Strähle V, Schütz G (1986). In vivo protein-DNA interactions in a glucocorticoid response element require the presence of the hormone. Nature 324:686-688.
69. Gaubert CM, Carriera R, Shyamala G (1986). Relationship between mammary estrogen receptor and estrogenic sensitivity. Molecular protperties of cytoplasmic receptor and its binding to deoxyribonucleic acid. Endocrinology 118:1504-1512.
70. Catelli MG, Ramachandran C, Gauthier Y, Leganeux V, Quelard C, Baulieu EE, Shyamala G (1988). Biochem J (submitted).

V. THERMOTOLERASE

Stress-Induced Proteins, pages 223–233

SURVIVAL OF CELLS EXPOSED TO ANTICANCER DRUGS AFTER STRESS[1]

George M. Hahn, Manohar K. Adwankar[2],
Vathsala S. Basrur, and Robin L. Anderson

Department of Radiation Oncology, Stanford University
Stanford, California 94305

ABSTRACT As hyperthermia becomes integrated into the clinic as a modality of cancer treatment, the possibility needs to be considered that cells surviving heat treatment are either temporarily or permanently resistant to anticancer drugs. Thermotolerant Chinese hamster cells are very resistant to adriamycin and partially resistant to bleomycin even at 37°C and pH 7.4. The pH dependence of cis-platinum cytotoxicity at elevated temperatures is nearly abolished in thermotolerant cells.
Experiments with RIF-1 murine tumor cells and their heat resistant variants have not shown any cross-resistance between currently used drugs and exposure to hyperthermic temperatures. There is, however, strong cross-resistance between heat and amphotericin B, a plasma membrane active drug that has been suggested as a useful member of multidrug therapy. This finding, coupled with the demonstration that amphotericin B-resistant cells are also heat resistant, strongly suggests that the plasma membrane is the initial "target" for thermal cell killing.

INTRODUCTION

The induction of the heat shock (or other stress)

[1]Supported by NIH Grants CA04542 and CA19386.
[2]Present address: Cancer Research Institute, Tata Memorial Center, Parel, Bombay 400 012, India.

response modifies, at least temporarily, many characteristics of mammalian cells. This may have considerable clinical importance, particularly in the treatment of cancer. The major point of interest to be considered in this paper is the possibility that cells, either during or after stress, are more resistant to anticancer treatments than are cells of the same phenotype that were not previously exposed to the stress inducer. This is clearly not just an academic problem. First, during a fractionated course of hyperthermia the induction of thermotolerance (i.e., transient heat resistance) greatly influences the timing of subsequent, individual treatments. Second, we have demonstrated that cells stressed by exposure to glucocorticoids (but not other steroids) become temporarily heat resistant (1). No data exist in the literature that examine the obviously more important question of whether or not glucocorticoid stress also induces temporary drug resistance. Many cancer patients are chronically on these hormones; obviously, it would be important to know if, as a consequence, their diseases responded to chemotherapy in an altered way. Subjeck et al. have suggested that anticancer agents themselves can induce synthesis of heat shock proteins (HSPs) (2). While their preliminary data suggest that the appearance of HSPs may not be associated with thermotolerance, nevertheless, the possibility exists that those cells that survive initial chemotherapy have altered responses to other drugs. Finally, there are now two preliminary reports that strongly suggest that X-irradiation is a potent inducer of gene amplification (3,4). Radiation therapy frequently precedes chemotherapy. This finding may offer one explanation to the frequently voiced observation by oncologists that recurrences of radiation-treated tumors rarely respond to chemotherapy.

A closely related topic is cross-resistance. Are cells that are genetically resistant to heat also resistant to specific drugs (and vice versa)? The heterogeneity of responses of cells from human malignancies makes it very likely that most tumors contain a fraction of heat resistant cells. Therefore, in those tumors, cells that survive heat treatment would also be resistant to drugs against which cross-resistance has been demonstrated, and offspring of these cells would have an unfavorable response to some chemotherapies. In addition, as clinical studies of the combination of drugs and heat become more frequent, the existence of cross-resistance could become progressively more important in deciding which drugs are to be used in this combined modality approach.

There is a final possible benefit that could be derived from studying cross-resistance between heat and drugs. In spite of an impressive effort by many investigators, we still do not know why mammalian cells die after (or during) exposure to moderate temperatures (41-45°C). If consistant cross-resistance could be established for one or more drugs, then very likely the "target" for cell killing would be similar. The mode of cytotoxicity and particularly the "target" for most drugs is reasonably well established. Cross-resistance, therefore, could go far in explaining how heat kills cells and, perhaps, how tolerance protects them.

METHODS AND MATERIALS

We have used two model systems to study these problems. The first is a well established line of Chinese hamster cells originally derived by HSU from ovaries; the particular line, HA1, was isolated by Yang et al. This system goes back to the 1950's (5). The second model is of very recent origin; it consists of heat resistant variants of a radiation induced fibrosarcoma (RIF-1) growing in C_3H mice (6). These variants (RIF-TR) were selected by repeated heat exposures, from cells not mutagenized; selection was based on the following criteria: 1. the cells all derive from a single parent line; 2. the cells show large difference in heat resistance from the parent cells; 3. *in vitro*, growth rates of all the cell strains do not vary appreciably from that of the parent line; 4. when injected into syngeneic hosts, cells from the strains give rise to tumors; 5. tumor growth rates of wild type and thermally resistant strains are similar; 6. there are no unusual morphological identifying characteristics among the cell strains; 7. cells from each of the cell strains grown *in vitro* give rise to colonies with good plating efficiences; 8. cells obtained directly from tumors, when dissociated and plated *in vitro*, also give rise to colonies, thus permitting precise assays of cellular responses *in vivo*; 9. at least some of the heat-induced cell strains should evoke only a minimal immune response when injected into syngeneic hosts.

RESULTS AND DISCUSSION

Effect of thermotolerance on HA1 cells

Adriamycin (ADM) This drug is one of the most frequently used anticancer agents. A time response of HA1 cells treated with a fixed dose of ADM at 43°C is shown in Figure 1A. As the response of these cells to hyperthermia alone is a function of cell density and of nutritive factors, all experiments were performed according to a common protocol described earlier (7). Figure 1A also

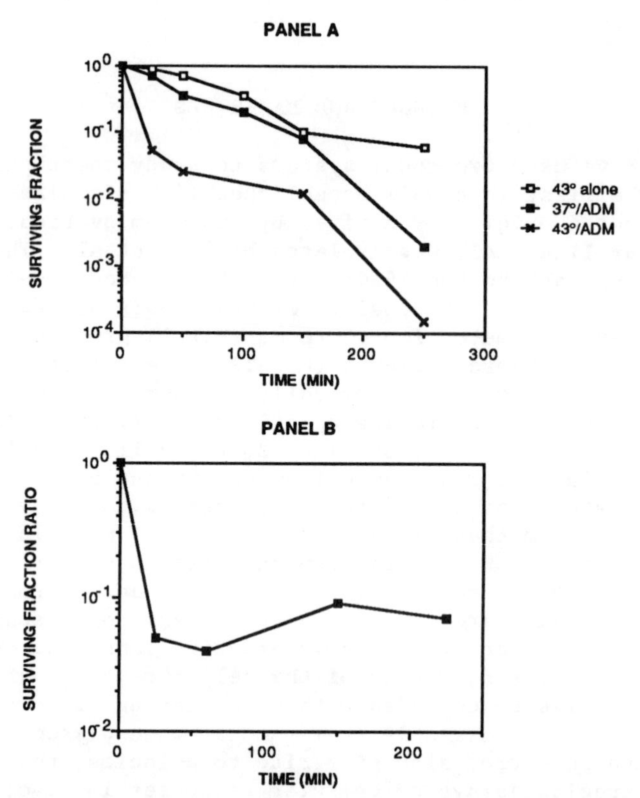

FIGURE 1--Time response of HA1 cells exposed simultaneously to 43°C and ADM. (A) Survival values. (B) Surviving fraction ratios which were obtained by dividing, point by point, the values of the lowest curve in (A) by the 43°C control values. The curve is a measure of the cytotoxic action of ADM at 43°C.

shows two controls: the survival of cells exposed to ADM at 37°C and that of cells heated to 43°C in the absence of ADM. The 37°C data showed that this survival curve was essentially exponential over the exposure time; the 43°C heat control data were consistent with data published many years ago (8). The combined curve showed appreciable cell killing for short-term exposures: At 30 min, survival was about $4x10^{-2}$. This contrasted with survival levels of 90% for the heat control and 60% for the ADM control. Hence, with independent action, a 54% level of survival would be predicted for the combined treatment. However, beyond the 30-minute point, the combined treatment curve approached the shape of the 43°C control; the former was essentially parallel to the latter, as illustrated in Figure 1B, which shows the effect of ADM at 43°C. This curve was obtained by normalizing the combined curve point by point to the heat control. Beyond about 30 min, ADM clearly no longer contributed to cell killing. In fact, there may have been protection against ADM by heat: The last two time points were higher in survival by a factor of 2 than the earlier points (though this difference is not statistically significant). These results were not due to thermal inactivation of ADM: Heating ADM to 43°C for up to 5 h did not appreciably reduce the drug's ability to subsequently kill cells at 37°C (data not shown). Thus, heat initially caused superadditive cell killing; the development of tolerance, however, was accompanied by protection. Later experiments (Fig. 2) showed that not only at 43°C but also at 37°C were thermotolerant cells protected against ADM. Cells remain ADM tolerant for at least three days (7).

In order to get an idea of the mechanism of this curious phenomenon, we monitored the incorporation of ADM into HA1 cells (9). This was done by monitoring the fluorescence of ADM with flow cytometry. Results showed that the mean log fluorescence was 18.9 for control cells (under the conditions described in 9), 22.7 for nontolerant cells exposed to ADM for 10 min at 45°C, while it was only 14.0 for cells preheated for 10 min at 45°C and 4 h later exposed to ADM. In addition, cell sorting experiments showed that for nonheated controls and for nontolerant cells exposed at 45°C, incorporation of ADM was inversely correlated with cell survival. For tolerant cells, however, increased ADM incorporation was not necessarily accompanied by reduced survival. Our data suggest two different mechanisms: a modulation of plasma membrane permeability to ADM by heat, resulting in increased uptake initially, followed by reduced uptake with the development of tolerance. In addition, the data indicate protection of the "target" (presumably DNA) against even those ADM molecules that had entered the thermotolerant cell.

Bleomycin (Bleo) Another drug that has shown efficacy against many malignancies is this glycopeptide. HA1 cells exposed to this drug become inactivated as shown in Fig 2A.

The temperature dependence of cell killing by Bleo behaves in a way characteristic for a class of agents: increased activity at elevated temperature with a threshold

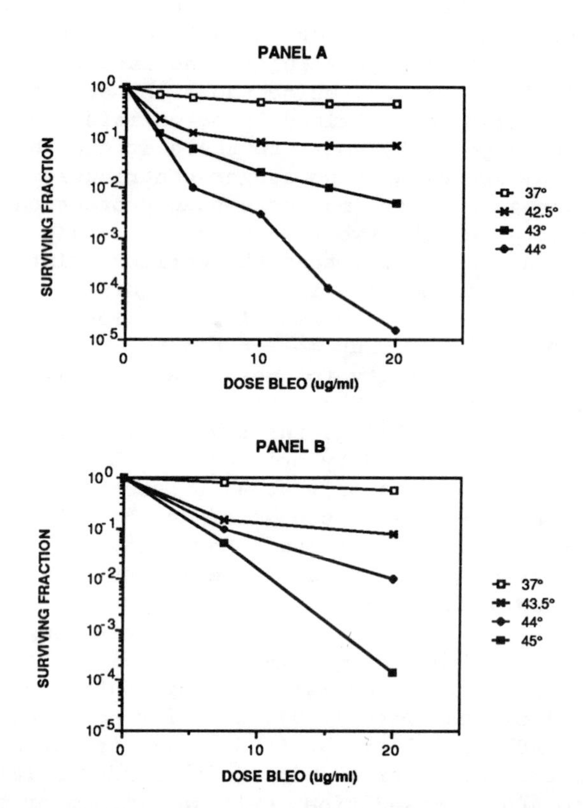

FIGURE 2--Response of previously unheated and thermotolerant HA1 cells to bleomycin. Panel A: Unheated cells; Panel B: Thermotolerant cells. While some protection is seen at all temperatures, the most striking aspect is the shift in the threshold temperature from 42.5°C (unheated cells) to 44°C for tolerant cells.

between 42 and 43°C. Below 42°C, temperature dependence is relatively minor; above the threshold it is quite dramatic. In thermotolerant cells, some protection is seen at all temperatures; the most striking aspect, however, is the change in threshold from 42 to 43.5°C (Fig 2B). Neither the reason for the threshold nor protection against Bleo by tolerance is currently understood.

Cis-platinum Perhaps the most effective antineoplastic drug today is cis-platinum. This drug, as a single agent, cures a high proportion of testicular tumors; in addition, it is effective against many other solid tumors. The cytotoxic efficacy of this agent at mildly elevated temperatures is greatly affected by the heat shock response. This is illustrated in Fig. 3 where cell killing to a fixed dose of this drug at 43°C (1 h exposure) is shown as a function of extracellular pH. pH dependence of cytotoxicity is of considerable interest in oncology because portions of many tumors are characterized by high proton density. In these tumor volumes, the pH takes on values between 6.5 and 7.0. By contrast, the pH of normal tissue rarely departs by more than a small fraction of a pH unit from the usual 7.4.

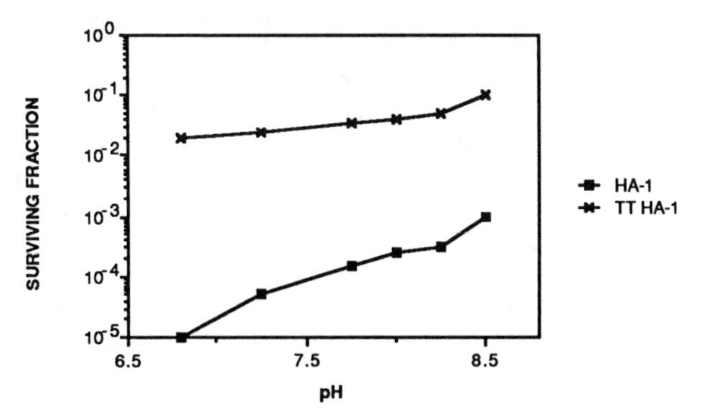

FIGURE 3--Effect of thermotolerance on pH dependence of cis-platinum killing. Cells were exposed to cis-ddp for 1 h at 43°C at the indicated pH. The thermotolerant cells were protected at all pH values; the most striking effect was noted at pH 7 or lower. Such low pH values are frequently associated with the interior of large solid tumors.

The data in Fig. 3 show clearly that at a pH of 7.0 or lower, cis-platinum is extremely efficient in killing HA1 cells at 43°C. If, however, the cells are exposed to a nonlethal heat dose 12 h before drug exposure, then the cytotoxicity is greatly reduced even at pH 7.4 or higher, and is no longer enhanced at low pH values. Again, no explanation exists that accounts for these findings.

Cross-resistance

We have examined several drugs for possible cross-resistance to heat, using the RIF-TR system (6). These include: the nitrosourea, BCNU, whose primary mode of cytotoxicity appears to be alkylation of DNA; Bleo, a drug that induces DNA strand breaks; and cis-platinum, whose mode of action very likely involves adding platinum adducts to DNA. There was no evidence of cross-resistance for any of these. Hence, these agents can be used on patients whose tumors had been treated with hyperthermia, without fear that heat may have selected for drug resistance. Furthermore, the results strongly reinforce the view that cell killing by heat does not involve DNA as the primary "target".

There was negligible cross-resistance between the anthracyclic antibiotic ADM and hyperthermia. At 37°C, and particularly at 43°C, the RIF-TR strains are somewhat more resistant to this drug than the parent line, RIF-1; a similar result was found for CHO cells by Wallner and Li (10). Until recently, it was thought that ADM killed cells exclusively by intercalating into DNA. Newer data, however, suggest an alternate mode of action. This involves, in an as yet unspecified way, the cells' plasma membranes (11). It is possible that this latter mode is enhanced at hyperthermic temperatures.

We are just beginning to examine the drugs (VP-16, VM-26) that have topoisomerase II as their "target". Our preliminary data suggest no cross-resistance for these; in fact, early data suggest collateral sensitivity: the heat resistant TR cells may be more sensitive, at least to VP-16, than the heat sensitive parent line.

Amphotericin B (AmB) This polyene antibiotic is not usually considered an anticancer agent, although several investigators have suggested it as a possible component of a multidrug "cocktail" (12). At the present time, it is employed clinically as an antifungal agent. We found that all TR lines are extremely resistant to this agent (Fig.4). Furthermore, RIF-1 cells made thermotolerant also became resistant to AmB. Because of this close correspondence between heat and AmB resistance, we selected AmB-resistant cells directly from RIF-1 cells by growing them in the presence of increasing concentrations of this antibiotic. These cells, termed RIF-AR, that had never been exposed to heat were, nevertheless, uniformly heat resistant. In other words, selection for AmB resistance implied heat resistance, while selection for heat resistance implied AmB resistance. Also, exposures of RIF-1 cells to AmB induced a modest amount of thermotolerance, while the thermotolerant cells displayed as much resistance to AmB at 43°C as RIF-AR.

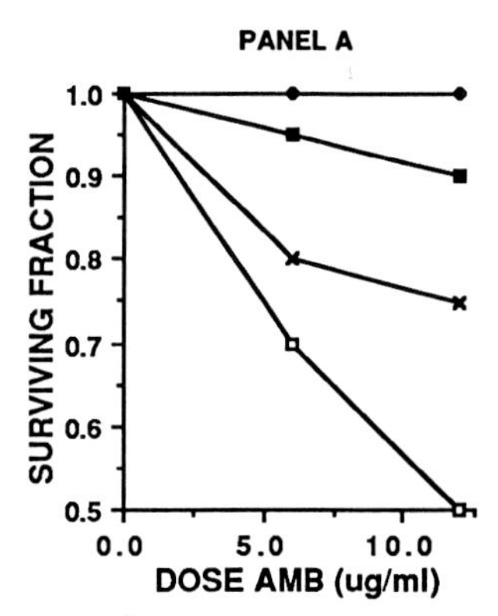

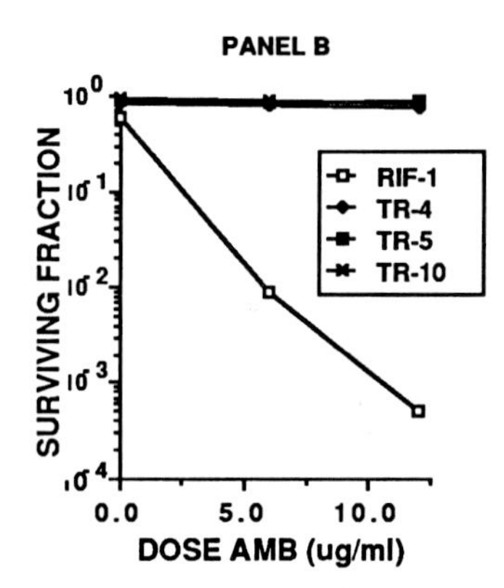

FIGURE 4--Resistance of TR strains to amphotericin B. Panel A: All the TR strains examined showed considerable resistance to this drug at 37°C when compared to the RIF-1 parent line. This effect was accentuated at 43°C (panel B).

The events leading to AmB-induced cytotoxicity have been investigated in considerable detail (13). The drug combines with cholesterol; cholesterol-AmB molecules aggregate into complexes that result in the formation of nonspecific plasma membrane ionophores. Presumably, too many ionophores puncture the membranes excessively, so that the cell can no longer maintain trans-membrane ion gradients. This situation rapidly leads to cell death. The major point that we want to emphasize is that the initial, and crucial events all take place within the plasma membrane. The close correspondence between heat-induced and AmB-induced events therefore strongly implies that for thermal-induced cell death there is a similar series of events also going on in the plasma membrane. Perhaps, the recent demonstration that heat induces intramembrane particles (i.e., protein aggregates) in the plasma membranes of heated cells provides a clue for the determination of the sequence of events leading to heat-induced cell death (14,15). The data on the response of heat-resistant variants, RIF-AR and thermotolerant RIF-1 cells to AmB at 43°C and induction of thermotolerance in RIF-1 cells by AmB suggest a common or similar mechanism of protection against these stresses.

ACKNOWLEDGMENTS

We thank Esther Shiu and Ine van Kersen for important technical contributions.

REFERENCES

1. Fisher, GA, Anderson, RL and Hahn, GM (1986). Glucocorticoid-induced heat resistance in mammalian cells. J Cell Physiol 128:127-132.
2. Hughes, CS, Cain, JW and Subjeck, JR (1988). Induction of heat shock proteins by anticancer agents. 36th Annual Radiation Research Society Meeting (Abstract).
3. Hahn, PJ, Nevaldine, BH, Sagerman, RH and King, GA (1988). Radiation induces MTX resistance. 36th Annual Radiat Res Soc Meeting (Abstract).
4. Hopwood, LH and Moulder, JE (1988). Enhancement of drug resistance following irradiation of RIF-1 and SCC VII tumors. 36th Annual Radiat Res Soc Meeting (Abstract).
5. Yang, SJ, Hahn, GM and Bagshaw, MA (1966). Chromosome aberrations induced by thymidine. Exp Cell Res 42:130-135.
6. Hahn, GM and van Kersen, I (1988). Isolation and initial characterization of thermoresistant RIF tumor cell strains. Cancer Res 48:1803-1807.
7. Hahn, GM and Strande, DP (1976). Cytotoxic effects of hyperthermia and adriamycin on Chinese hamster cells. J Natl Cancer Inst 57:5:1063-1067.
8. Hahn, GM (1974). Metabolic aspects of the role of hyperthermia in mammalian cell inactivation and their possible relevance to cancer treatment. Cancer Res 34:3117-3123.

9. Rice, GC and Hahn, GM (1987). Modulation of adriamycin transport by hyperthermia as measured by fluorescence-activated cell sorting. Cancer Chemother Pharmacol 20:183-187.
10. Wallner, K and Li, GC (1986). Adriamycin resistance, heat resistance and radiation response in Chinese hamster fibroblasts. Int. J. Radiation Oncology Biol. Phys. 12:829-833.
11. Tritton, TR and Yee, G (1982). The anticancer agent adriamycin can be actively cytotoxic without entering cells. Science 217:248-251.
12. Medoff, G, Schlessinger, D and Kobayashi, GS (1973). Polyene potentiation of antitumor agents. J Natl Canc Inst 50:1047-1050.
13. Demel, RA, van Deenen, LLM and Kinsky, SC (1965). Penetration of lipid monolayers by polyene antibiotics. J Biol Chem 240:2749-2753.
14. Arancia, G, Malorni, W, Mariutti, G and Trovalusci, P (1986). Effect of hyperthermia on the plasma membrane structure of Chinese hamster v 79 fibroblasts: a quantitative freeze fracture study. Radiat Res 106:47-55.
15. Rice, GC, Fisher, KA, Fisher, GA and Hahn, GM (1987). Correlation of mammalian cell killing by heat shock to intramembranous particle aggregation and lateral phase separation using fluorescence-activated cell sorting. Radiat Res 112:351-364.

Stress-Induced Proteins, pages 235–244

THE *FORKED* PHENOCOPY IS PREVENTED IN THERMOTOLERANT PUPAE

Nancy S. Petersen and Herschel K. Mitchell
Department of Molecular Biology, University of Wyoming, Laramie, Wyoming 82071
and
Biology Division, California Institute of Technology, Pasadena, California 91125

ABSTRACT The same treatment which induces tolerance to heat and cold in *Drosophila* larvae also can prevent developmental defects induced by heat in pupae. The mechanism by which the defects, called phenocopies, are prevented is being studied in heterozygotes of the recessive mutant *forked*. The *forked* phenotype can be induced in females heterozygous for the mutation (f/+) but not in wild type flies. Here we show that the same thermotolerance inducing treatment that prevents wild type phenocopies also prevents induction of the *forked* phenotype in mutant heterozygotes.

INTRODUCTION

Thermotolerance can be induced in a wide variety of systems by a less extreme heat treatment prior to a normally lethal heat shock (1-3). In *Drosophila* larvae, the same treatment that induces thermotolerance, a 30 minute exposure to 35°C, can induce tolerance to cold, 0°C treatment (4). The 35° treatment changes cells so that many processes including ribosome synthesis, mRNA synthesis and processing, cytoskeletal organization, and protein synthesis recover much faster in the thermotolerant state (5-7). Because of the difficulty in determining which of these effects is responsible for improved survival, we have chosen to look at the effects of the 35° treatment on the prevention of developmental defects.

A wide variety of developmental defects can be induced by heating *Drosophila* pupae at 40.2-40.8°C for 30 to 40 minutes(8). These developmental defects are induced in 100% of the treated animals and they are called phenocopies because of their resemblance to mutant phenotypes. Heat induced phenocopies are thought to result from the failure to synthesize a specific gene products required for normal development. Each phenocopy can only be induced during a specific sensitive period which appears to correspond to the time of expression of the gene product(9).

All of the phenocopies studied so far can be prevented if pupae are made thermotolerant by heating at 35°C for 30 minutes before the high temperature shock (10-12). This implies that there is a common mechanism for both the induction and the prevention of these developmental defects.

We are studying the molecular basis for the induction of phenocopies in sensitive animals and their prevention in thermotolerant animals. We can do this because we have been able to induce a phenocopy of the recessive mutant *forked*. The bristle phenocopy can be induced by heating *forked/+* pupae at 32-33 hours after puparium formation (13). Since the gene for *forked* has been identified and sequenced, the door is now open to studying how heat affects the expression of this gene (14,15). In this paper we show that the *forked* phenocopy can be prevented by a 30 minute 35°C heat shock in a similar manner to phenocopies induced in wild type flies. We will discuss how the phenocopies may be induced and which effects of the 35° C treatment might be important for phenocopy prevention.

RESULTS

A developmental defect in adult *Drosophila* bristles which closely resembles the *forked* phenotype can be induced in *forked/+* heterozygotes by heating the flies during pupal development (13).

All pupae heated to 40.7°C for 30 minutes between 32 and 33 hours of pupal development have many extremely *forked* bristles when they emerge as adults. Pupae heated slightly earlier or later show mild bristle defects, and pupae heated before 31 hours or after 34 hours resemble the wild type (see Figure 1). *Forked* bristles are induced over a fairly large region of the fly including the head and thorax by a single heat shock. This indicates that the sensitive periods for induction of this phenocopy are fairly broad and overlap for different tissues as compared to the multihair phenocopy which has quite discrete sensitive periods on different parts of the animal (16).

The effect of the 35°C treatment on *forked* phenocopy induction is dramatic. Figure 2 shows that both the number of bristles affected and the severity of the phenotype are reduced in animals which receive the protecting treatment immediately before the 40.2°C shock. This effect is quantitated in Figure 3 which shows the average number of affected bristles per fly in tolerant and non-tolerant flies which had been heated at either 32 or 33 hours after puparium formation. The conditions used for phenocopy induction are described in the legend and in reference (13). Heat shocks at both 32 and 33 hours after puparium formation are used to show that the protecting effect is not due to a shift in developmental stage induced by the 35°C treatment. The 35° treatment actually slows down the developmental program of protein synthesis, but if this were the reason for not getting defects at 32 hours it certainly cannot be the reason for not getting the defect induced at 33 hours where the animals are clearly well into the sensitive period before the heat shock. Conversely, if the heat shock speeds up some other process important for phenocopy induction, a half hour heat shock would have to change the developmental timing by more than an hour to account for the effects observed.

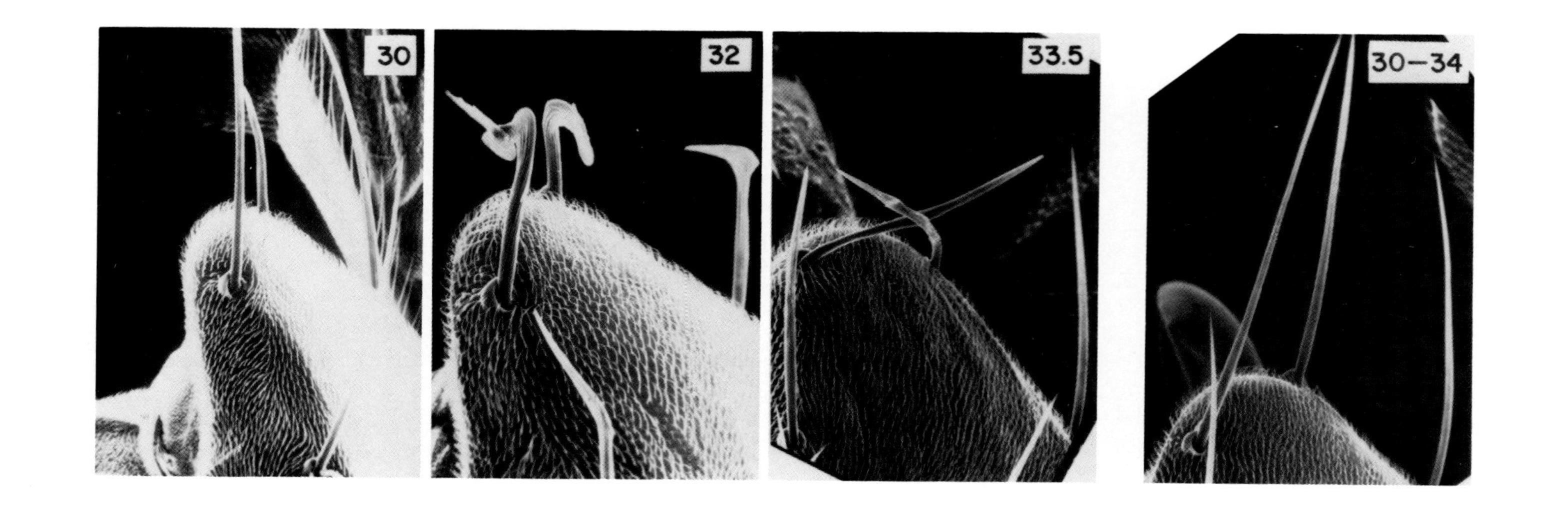

FIGURE 1. The forked phenocopy induced in female heterozygotes (f/+). Pupae were heated for 30 minutes at 40.7°C at the developmental times indicated. The pictures are of the scutellar bristles of emerged adults. Figure from (13).

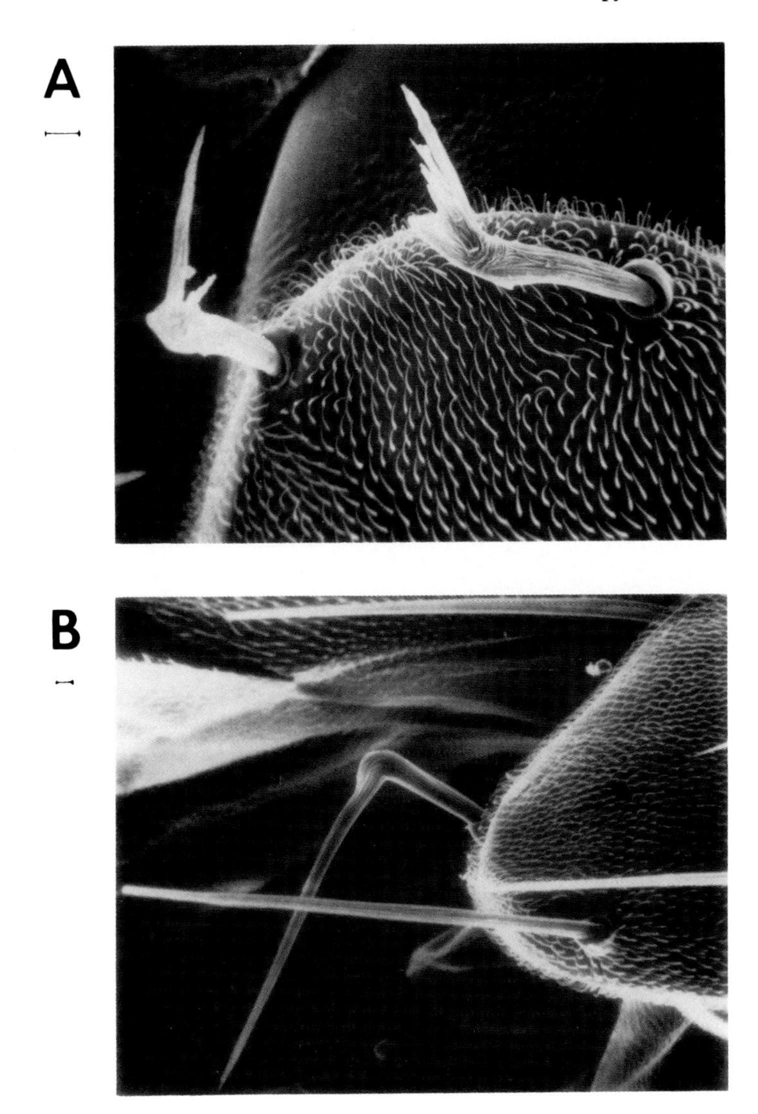

FIGURE 2. Both the number and the severity of the *forked* phenocopy is reduced when pupae are made thermotolerant before the phenocopy inducing heat shock. **A**. Scutellar bristles of an adult f/+ fly which was heat shocked for 30 minutes at 40.2°C at 32 hours of pupal development. **B**. Scutellar bristles of a fly which received the same treatment as in A, but was heated to 35°C for 30 minutes immediately prior to the high temperature shock.

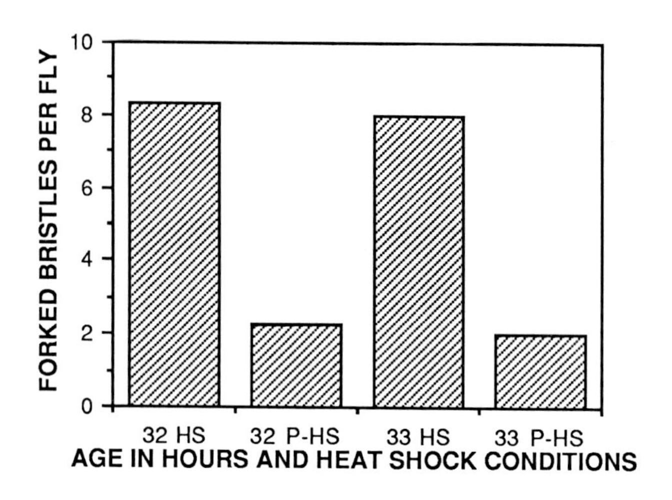

FIGURE 3. The number of *forked* bristles per fly is markedly reduced when pupae are given a thermotolerance inducing treatment before the phenocopy inducing heat shock. 32 HS indicates the number of *forked* bristles per fly when f/+ pupae are heated for 30 minutes at 40.2°C at 32 hours after puparium formation. 32 P-HS indicates that pupae were first heated at 35°C between 31.5 and 32 hours of development and then given a 40.2°C heat shock for 30 minutes starting at 32 hours. The corresponding treatments were repeated an hour later on 33 hour pupae indicated as 33HS and 33P-HS.

DISCUSSION

Heat induced defects are probably due to the failure to synthesize specific gene products at the correct time in development (9-11). The induction of mutant phenotypes in heterozygotes by heat shock lends support to this point of view. Figure 4 shows how we envision the effects of heat acting on the *forked* gene product. The underlying assumption is that a certain amount of the normal or wild type gene product must be made during the sensitive period indicated by dark vertical lines in order for bristles to develop normally. In the mutant heterozygote the level of normal gene expression is just above the minimum required for normal development. The heat shock alters either the level (**A**) or the timing (**B**) of gene expression so that the required amount of gene product is not present during the sensitive period. This can be done either by an absolute decrease in the amount of gene product made or by a changing the timing of synthesis so that the product is not made at the correct time in development. In reality some combination of these two theoretical cases may be correct.

In order to actually determine how the *forked* phenocopy is induced it will be necessary to determine the amounts of the *forked* mRNA and protein made under these different conditions. These experiments are in progress.

The prevention of the *forked* phenocopy by the 35°C heat treatment has several important implications. First, it implies that heterozygote phenocopies are induced by the same mechanism as phenocopies in wild type flies. Secondly, it suggests that both phenocopy induction and the protective effects of the thermotolerance inducing treatment act through effects on gene expression during the phenocopy sensitive period.

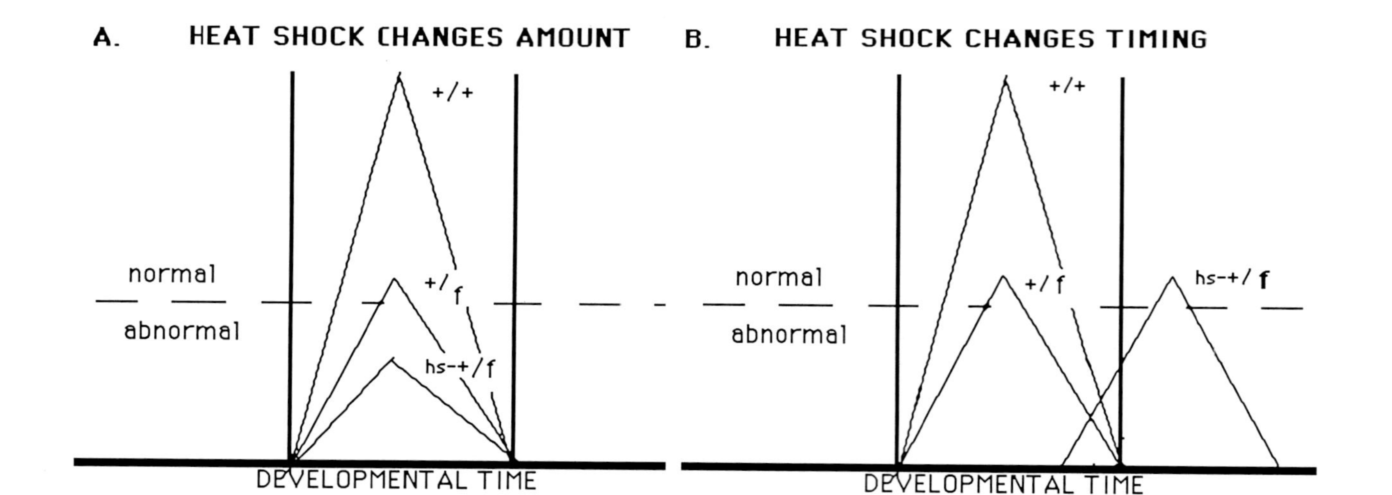

FIGURE 4. Hypothetical levels of forked gene expression in pupae with different genotypes. The heights of the diagonal lines represent the amount of gene product synthesized at different stages in development by +/+ wild type pupae; +/f forked heterozygote pupae; and hs-+/f by heat shocked heterozygote pupae. The horizontal dashed line represents the level of gene product required to get normal bristles. A. Heat shock changes the amount of forked gene product made. B. Heat shock changes the timing of synthesis of the forked gene product.

REFERENCES

1. Gerner EW, Schneider MJ (1975). Induced thermal resistance in HeLa cells. Nature(London) 256:500.
2. Mitchell HK, Moller G, Petersen NS, Lipps-Sarmiento L, (1979). Specific protection from phenocopy induction by heat shock. Dev. Genetics 1:181.
3. Schlesinger MJ, Ashburner M, Tissieres A, (1982). " Heat shock from bacteria to man." Cold Spring Harbor NY , Cold Spring Harbor Press.
4. Burton V, Mitchell HK., Young P, Petersen NS, (1988). Heat shock protection against coldstress in *Drosophila melanogaster*. Mol. Cell. Biol. in press.
5. Craig EA, (1985). The heat shock response. CRC Critical Reviews in Biochemistry. 18:239.
6. Lindquist S, (1986). The heat-shock response. Ann. Rev. Biochem. 55:1151.
7. Neidhardt FC, VanBoggelen RA, Vaughn V, (1984). The genetics and regulation of heat shock proteins. Ann. Rev. Genet. 18:295.
8. Mitchell HK, Petersen NS, (1982). Developmental abnormalities induced by heat shock. Dev. Genet. 3:91.
9. Mitchell HK, Lipps LS (1978). Heat shock and phenocopy induction in Drosophila. Cell 15:907.
10. Petersen NS, Mitchell HK (1982). Induction and prevention of the multihair phenocopy in Drosophila, In Schlessinger M, Ashburner M, Tissieres A (eds): "Heat shock from bacteria to man," New York: Cold Spring Harbor Press. p345.
11. Petersen NS, Mitchell HK (1985). Heat shock proteins. In Kerkut GA, Gilbert LI (eds): " Comprehensive Insect Physiology, Biochemistry, and Pharmacology, Vol.X. Biochemistry," Oxford: Pergamon Press, p 347.
12. Petersen NS, Mitchell HK (1987) The induction of a multiple wing hair phenocopy by heat shock in mutant heterozygotes. Dev. Biol. 121:335.

13. Mitchell HK, Petersen NS, (1985). The Recessive phenotype of forked can be uncovered by heat shock in *Drosophila*. Develop. Genet. 6:93.
14. Parkhurst SM, Corces VG (1985). *forked*, gypsys, and suppressors in *Drosophila*.Cell 41:429.
15. McLaughlin A, (1986). *Drosophila* forked locus. Mol. Cell. Biol. 6:1.
16. Mitchell HK, Petersen NS, (1981). Rapid changes in gene expression in differentiating tissues of Drosophila. Dev. Biol. 85:233.

VI. CLINICAL PROBLEMS

Stress-Induced Proteins, pages 247–256

BIOCHEMICAL AND CELLULAR RESPONSES TO HYPERTHERMIA IN CANCER THERAPY[1]

Stephen W. Carper, Paul M. Harari,
David J.M. Fuller, and Eugene W. Gerner

Department of Radiation Oncology and Cancer Center
University of Arizona Health Sciences Center
Tucson, Arizona 85724

ABSTRACT Hyperthermia is used clinically to treat certain types of cancer in humans. Increasing cellular temperature can effect the biochemistry and physiology of the cell in ways that are complex and not fully understood. One specific effect of heat shock is to induce polyamine catabolism. Thermal stress-induced polyamine acetylation, the first step in the catabolic pathway, was found to occur in both prokaryotes and eukaryotes. Spermidine acetyltransferase (SAT), the enzyme responsible for the acetylation of spermidine was induced by heat in all human tumor cell lines investigated, but to varying degrees. The accumulation of N^1acetylspermidine was observed in most human cell lines that showed SAT induction. In mammalian cells, N^1acetylspermidine is converted to putrescine by polyamine oxidase (PAO), with the subsequent production of H_2O_2 and 3-acetamidopropanal. The H_2O_2 generated by PAO from the heat-induced N^1acetylspermidine was not toxic to Chinese hamster ovary cells (CHO). Inhibitor studies indicated that CHO cells apparently did not detoxify the H_2O_2 by glutathione oxidation or by catalase. SAT may have a physiological role in the maintenance or decay of the thermotolerant state in cells.

[1]This work was supported in part by grants from the USPHS, National Cancer Institute CA-30052, CA-17343 and CA-09213.

INTRODUCTION

Hyperthermia, the elevation of cellular temperature above the normal growth temperature, has a plethora of effects. For mammalian cells growing at 37°C, an elevation of temperature up to or above 42°C is toxic. This cytotoxicity has been exploited clinically for the treatment of certain types of cancer (1,2,3). Cancer cells are not inherently more thermosensitive than their normal counterparts; however, the physical environment of the tumor (i.e., low pH and decreased nutrient availability) apparently increases the cytotoxic activity of a heat shock. Clinical applications of hyperthermia are most often administered in combination with radiation or chemotherapy due to a synergistic interaction of heat with radiation or drugs (4). Besides being cytotoxic, thermal stress induces the synthesis of a unique set of proteins known as heat shock proteins (hsp)(for reviews see 5,6,7). If cells are heated, allowed to recover for a few hours at their normal growth temperature, and then reheated, they become resistent to heat induced cytotoxicity. This transient, nonheritable phenomenon has been termed thermotolerance. The development of thermotolerance occurs at about the same time as the synthesis of hsp. This has prompted some researchers to suggest that hsps are involved in the development of thermotolerance. However, a careful review of the literature shows that thermotolerance is a complex process, and that the possible role of hsps in protecting the cell from heat damage or repairing heat damage is still poorly defined (8).

The molecular mechanism by which heat kills cells is not understood. Hyperthermia affects a wide variety of cellular structures and functions, and causes changes in mitochondria and nuclear structure, denaturation of soluble and membrane proteins as well as a disassembly of actin microfilament bundles. These structures will reform or refold at normal growth temperatures. Heat shock also effects major biochemical pathways. The polyamine biosynthetic pathway is dramatically altered by hyperthermic treatment. Polyamines are naturally occurring polycations, found in high concentrations (near millimolar), that are required for a variety of cellular processes including growth and differentiation (9). Our laboratory has shown that ornithine decarboxylase, the first enzyme in the polyamine pathway, is inhibited by heat shock (10).

Recently we have shown that one of the enzymes involved in polyamine catabolism, spermidine/ spermine N^1-acetyltransferase (SAT) is heat inducible in Chinese hamster ovary (CHO) cells (11).

In order to determine whether the effect of heat shock on polyamine catabolism was unique to hamster cells, we investigated the acetylation of polyamines in prokaryotic and eukaryotic cell lines. The ability of cold shock to induce polyamine acetylation and the effects of heat induced polyamine oxidation on cell viability in CHO cells were also studied.

MATERIALS AND METHODS

Cell Culture

All mammalian cells were cultured in McCoy's 5A medium supplemented with 10% fetal bovine serum and maintained at 37°C in humidified incubators with 5% CO_2:95 air. *E. coli* (strain LE 392) were cultured in LB medium with vigorous shaking at 37°C.

Polyamine Analysis

Polyamines were separated and analyzed by reverse phase high performance liquid chromotography as previously described (12). Protein was determined by the method of Bradford (13).

Spermidine N^1Acetyltransferase Activity

Enzyme activity was measured by incorporation of ^{14}C label into spermidine from [^{14}C] acetyl-CoA as previously described (14).

Survival Determinations

After heat shock, single cells were plated into 60 mm petri dishes, incubated for 7-10 days, stained with crystal violet and counted if colonies contained more than 50 cells as previously described (15).

Drugs

Buthionine S,R sulfoximine (BSO) and aminotriazole were purchased from Sigma Chemical Co. [1-^{14}C]-acetyl-

CoA was purchased from New England Nuclear. The polyamine oxidase inhibitor, MDL-72.521, was a gift from Merrell Dow Laboratories, Cincinnati, Ohio.

RESULTS

Temperature stresses can induce changes in polyamine metabolism in E. coli. Figure 1 shows that there is a rapid accumulation of both N^1acetylspermidine and N^8acetylspermidine after actively growing cultures are placed into an ice water bath.

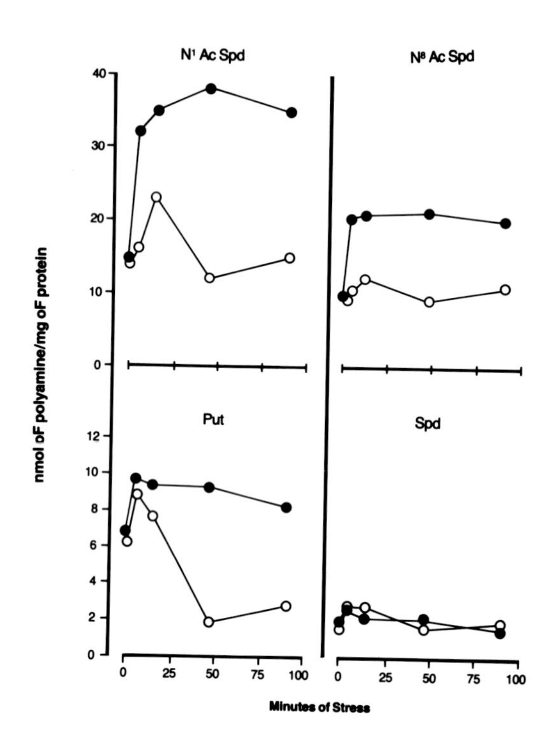

Figure 1. Polyamine levels following temperature stresses in E. coli. The filled symbols are for cultures exposed to 0.4°C, while the open symbols represent cultures exposed to 42°C continuously for the times indicated.

Increasing the culture temperature from 37°C to 43°C also results in an elevation of N^1 and N^8acetylspermidine levels. However, the heat shock levels of N^1 and N^8acetyl-spermidine were approximately half of those obtained by cold shock. Cold shock resulted not only in a more rapid increase in the contents of these polyamines but once formed, the levels of the acetylated polyamines remained at the higher values longer, probably due to a decrease in metabolism at the colder temperature. These results clearly indicate that prokaryotes alter their polyamine levels during extreme temperature stresses.

Figure 2 shows that a hyperthermic stress can induce SAT activity in a variety of human tumor cell lines.

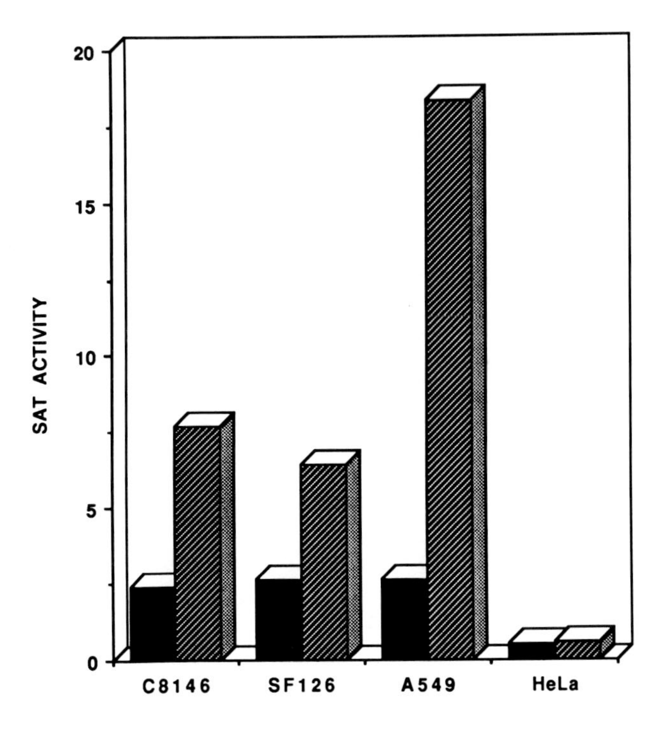

Figure 2. Heat inducible spermidine acetyltransferase activity in human tumor cell lines. SAT activity (pmol/min/10^6 cells) in control (dark bars) or 6 hrs after a 45°C, 20 min heat shock (hatched bars).

The SAT activity in cultures growing at 37°C increases as cells reach plateau phase (unpublished results). The A549 cells (human lung cancer, ATCC #CCL185) display the largest increase in SAT activity 6 hours after a 45°C, 20 minute heat shock, while HeLa cells (human cervical cancer) showed little or no SAT induction following this heat shock. Chemical stress (diethyldithiocarbamate) can also induce SAT activity in A549 cells but not in HeLa cells, while cold shock did not induce SAT activity in A549 cells (data not shown).

The levels of N^1acetylspermidine, the product of SAT, are reported in Table 1.

TABLE 1
LEVELS OF N^1 ACETYLSPERMIDINE IN HUMAN TUMOR CELL LINES

Cell Line	N^1acetylspermidine Levels (nmol/mg protein) Control	6 Hrs. After a 45°C, 20 Min. Heat Shock	
	PAOI (-)	PAOI (-)	PAOI (+)
C8146	0.15	0.63	0.56
SF-126	n.d.	n.d.	n.d.
A549	0.20	0.37	1.68
HeLa	n.d.	n.d.	n.d.

PAOI = Polyamine oxidase inhibitor
n.d. = not detectable

The polyamine oxidase inhibitor (PAOI), MDL 75.521, was added to prevent the conversion of N^1acetylspermidine to putrescine. The A549 cells show the largest accumulation of N^1acetylspermidine while HeLa and SF-126 cells (brain tumor cells) show no N^1acetylspermidine production. The lack of accumulation of N^1acetylspermidine in SF-126 cells is puzzling since these cells induce SAT activity following a heat shock. The PAOI had no effect on the accumulation of N^1acetylspermidine in C8146 cells (melanoma cells) while a dramatic increase was observed in the A549 cells. These differences are probably due to variations in the activity of the polyamine oxidase in these different cell lines (i.e. the polyamine oxidase is more active in A549 cells, hence a greater accumulation of substrate [N^1acetylspermidine] is seen in the presence of the PAOI).

To investigate the possibility that the H_2O_2 produced by PAO following a heat shock is toxic to cells, the following experiments were performed. CHO cells were cultured in the presence or absence of 25 μM PAOI and heat shocked at 43°C. Figure 3A shows that inhibition of the PAO did not result in an increase in cell survival following a heat shock.

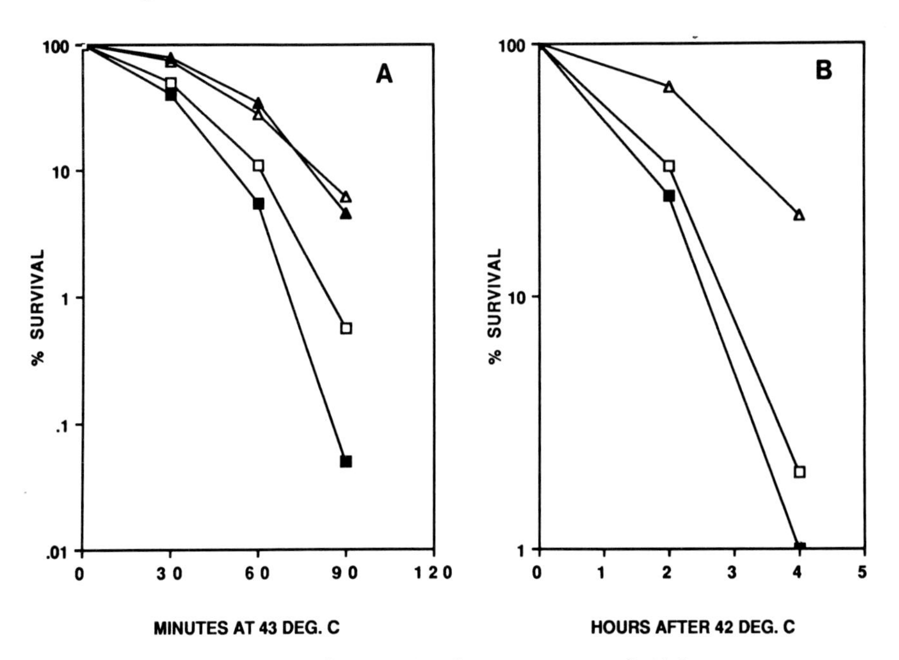

Figure 3. Thermal survival response of Chinese hamster ovary cells. Panel A: cells were treated with 25 μM polyamine oxidase inhibitor (PAOI, MDL 75.521) two hours prior to heating (▲) or with 10 μM BSO added 24 hr prior to heating (■) or with both PAOI and BSO ([]). Panel B: cells were treated with 20 mM aminotriazole 2 hr prior to heating (■) or with 25 μM PAOI and 20 mM aminotriazole 2 hr prior to heating ([]). In both panels the drugs were present during heating. Non-drug treated cells are represented by (Δ).

Therefore, we concluded that the CHO cells were able to detoxify H_2O_2 formed as a result of heat shock. The two most common methods for H_2O_2 detoxification are catalase and glutathione oxidation. However, when glutathione levels were depleted by 70% with the inhibitor buthionine sulfoximine (BSO) (data not shown) or catalase activity was inhibited with aminotriazole (Figure 3B), the PAOI had little effect on increasing cell survival following heat shock. Both inhibitors did increase the sensitivity of CHO cells to heat through some mechanism which appears independent of polyamine oxidation.

DISCUSSION

The data presented in Figures 1 and 2 and Table 1 demonstrate that heat shock induces polyamine, and especially spermidine, acetylation. This phenomenon is observed in prokaryotes as well as in eukaryotes. However, cell line differences do exist in the amount and extent of the acetylation. Figure 3 shows that oxidation of N^1acetylspermidine to putrescine, 3-acetamidopropanal and H_2O_2 is not toxic, to CHO cells, suggesting H_2O_2 must be detoxified by some endogenous mechanism(s). Inhibitor studies indicate that the H_2O_2 detoxification does not occur by glutathione reduction or by catalase in CHO cells.

The physiological consequences of polyamine oxidation following a heat shock remain unclear. Perhaps the cell needs to remove or convert the acetylCoA that is released from the mitochondria following a heat shock. The formation of N^1acetylspermidine from acetylCoA and spermidine could effectively convert large amounts of acetylCoA into CoA. It is interesting to note that in HeLa cells which show a rapid decay of thermotolerance (15), there is little SAT induction. Roizin-Towle et al., have recently shown that depleting A549 cells of putrescine and spermidine by using α-difluoromethylornithine (DFMO, an inhibitor of ornithine decarboxylase) results in a rapid decay of thermotolerance (1988 Annual Meeting of the Radiation Research Society, Abstr. #Bi-6). We shave shown that treating CHO cells with DFMO inhibits the induction of SAT following a heat shock (D.J.M. Fuller, manuscript in preparation). Perhaps SAT plays an important role in the maintenance and decay of thermotolerance.

ACKNOWLEDGEMENTS

The authors wish to thank Sally Anderson for preparation of the manuscript.

REFERENCES

1. Wizenberg, M.J., Robinson, J.E. (1976). "Proceedings of the International Symposium on Cancer Therapy by Hyperthermia and Radiation." Philadelphia: ACR Press.
2. Storm, F.K. (1983). "Hyperthermia in Cancer Therapy." Boston: G.K. Hall Medical Publishers.
3. Overgaard, J. (1984). "Hyperthermic Oncology Vols. 1 and 2." London: Taylor and Francis.
4. Dethlefsen, L.A., Dewey, W.C. (1982). Third International Symposium: Cancer Therapy by Hyeprthermia, Drugs and Radiation. National Cancer Institute Monograph 61:1.
5. Craig, E.A. (1985). The heat shock response. CRC Crit Rev Biochem 18:329.
6. Lindquist, S. (1986). The heat shock repsonse. Annu Rev Biochem 55:1151.
7. Subjeck, J.R., Shyy, T.T. (1986). Stress protein systems of mammalian cells. Am J Physiol 250:c1.
8. Carper, S.W., Duffy, J.J., Gerner, E.W. (1987). Heat shock proteins in thermotolerance and other cellular processes. Cancer Res 47:5249.
9. Pegg, A.E. (1986). Recent advances in the biochemistry of polyamines in eukaryotes. Biochem J 234:249.
10. Gerner, E.W., Stickney, D.G., Herman, T.S., Fuller, D.J.M. (1983). Polyamines and polyamine biosynthesis in cells exposed to hyperthermia. Radiat Res 93:340.
11. Harari, P.M., Tome, M.E., Fuller, D.J.M., Carper, S.W., Gerner, E.W. (1988). Effects of diethyldithiocarbamate and endogenous polyamine content on cellular responses to hydrogen peroxide cytotoxicity. Int J Radiat Oncol Biol Phys (in press).
12. Seiler, N., Knodgen, B. (1980). High-performance liquid chromatographic procedure for the simultaneous determination fo the natural polyamines and their monoacetyl derivatives. J Chromat 221:227.

13. Bradford, M.M. (1976). A rapid and sensitive method for the quantitation of microgram quantitives of protein utilizing the principle of protein-dye binding. Anal Biochem 72:248.
14. Persson, L. Pegg, A.E. (1984). Studies of the induction of spermidine/spermine N^1-acetyltransferase using a specific antiserum. J Biol Chem 259:12364.
15. Gerner, E.W., Boone, R., Connor, W.G., Hicks, J.A., Boone, M.L.M. (1976). A transient thermotolerant survival response produced by single thermal doses in HeLa cells. Cancer Res 36:1035.
16. Starke, P.E., Farber, J.L. (1985). Endogenous defenses against the cytotoxicity of hydrogen peroxide in cultured rat hepatocytes. J Biol Chem 260:86.

Stress-Induced Proteins, pages 257–274

DISTINCT MEDIATORS AND MECHANISMS REGULATE HUMAN ACUTE PHASE GENE EXPRESSION

David H. Perlmutter, M.D.

Department of Pediatrics, Cell Biology and Physiology
Washington University School of Medicine
St.Louis, MO 63110

ABSTRACT Fever and profound changes in multiple metabolic pathways accompany the host response to inflammation/tissue injury. This 'acute phase response' is thought to result from activation of local tissue and circulating blood mononuclear cells with consequent release of several specific cytokines. Accordingly, we examined the effect of these several cytokines on expression of acute phase plasma proteins in human hepatoma cells (HepG2, Hep3B) and in primary cultures of human monocytes and macrophages. Recombinant human IL-1 increases steady state levels of mRNA and synthesis of complement proteins C3, factor B, α-1-antichymotrypsin and decreases steady state levels of mRNA and synthesis of albumin and transferrin. Preliminary results suggest that a specific ligand binding domain of IL-1 beta is, at least in part, responsible for the effect of that monokine on hepatic factor B gene expression. Deletion analysis of the upstream flanking region of the factor B gene indicates that specific sequence elements in this region are involved in IL-1 regulation. Hepatic acute phase gene expression is also regulated by the direct and distinct effects of monokines cachectin/ tumor necrosis factor and interferon-beta 2. Acute phase gene expression is elicited by both soluble and cell-associated monokines and may be activated both locally, in macrophages, and remotely, in liver. Several indirect pathways triggered by the actions of lymphokines, such as IL-2 and INF-γ, also result in acute phase gene expression. Individual acute phase genes may be regulated by several different monokines while other acute phase genes are affected only by a single monokine. For example, expression of factor B is regulated by the distinct effect of IL-1 beta, TNF alpha, INF beta 2, and INF-gamma, but expression of α_1 proteinase inhibitor is only affected by INF beta 2. Finally, extrahepatic sites of synthesis for acute phase proteins may add an additional level of complexity to this remarkable adaptive mechanism.

INTRODUCTION

During the host response to inflammation or tissue injury, there is a coordinate series of metabolic reactions that constitute the 'acute phase response'. These reactions include fever, muscle proteolysis, leukocytosis, alteration in fat, carbohydrate and trace mineral metabolism and profound changes in plasma concentrations of many liver-derived glycoproteins known as the acute phase plasma proteins, or acute phase reactants.

The potential significance of this response was recognized over 50 years ago when C-reactive protein was found in the sera of individuals during the acute stages of pneumococcal pneumonia and disappeared during convalescence as type-specific antibodies appeared (1). Plasma concentrations of some acute phase reactants increase (C-reactive protein, serum amyloid A, fibrinogen, α_1 proteinase inhibitor, complement proteins factor B and C3) while others decrease (albumin, transferrin). The changes in absolute concentration are now known to be largely caused by changes in rates of synthesis. Thus, this acute phase response involves a sequential and orderly activation of multiple genes. Recent studies of the acute phase response, as described below, have led to the identification of several distinct mediators, distinct pathways for activation and for expression of the responding genes.

There are several obvious similarities between the acute phase response and the stress response. A response to inflammation/tissue injury could, in fact, be viewed as a response to stress. Both systems involve relatively rapid activation of a group of genes highly conserved in evolution. Furthermore, sequence elements homologous with the consensus heat shock promoter have been identified in the upstream noncoding regions of acute phase genes C-reactive protein (2) and complement protein factor B (3). Nevertheless, there is no evidence that these elements are used to regulate these genes. Furthermore, there is no evidence for induction of the traditional acute phase reactants during heat shock or induction of the so-called heat shock/stress proteins during acute inflammation. Finally, it appears that the majority of the acute phase reactants are secreted proteins whereas the heat shock/stress proteins are localized to intracellular sites.

In the following paper I will review recent studies of the human acute phase plasma proteins. Due to limitations in the availability and long-term cultivation of human hepatocytes it has been difficult to study the response of human liver during acute inflammation. However, the acute phase response has been recently studied in cell lines [human hepatomas HepG2, Hep3B (4,5)] and in primary cultures [blood monocytes, tissue macrophages (6,7) and fibroblasts (8)] using recombinant mediators.

Mediators

The mediators of the host response to inflammation are soluble products of activated monocytes and/or lymphocytes. Interleukin-1 (IL-1) was the first of these mediators to be characterized structurally and functionally (reviewed in 9). There are at least two IL-1s, IL-1 alpha and IL-1 beta (10). IL-1 alpha, pI 5.0, is the predominant IL-1 in mice and IL-1 beta, pI 7.0, is the predominant IL-1 in man. These interleukins are encoded by distinct genes and bear only 25-30% overall primary structural homology. Nevertheless, the two IL-1s are recognized by the same cellular receptor and have similar biological activity in multiple tissues (9). Synthesis of the IL-1 alpha and IL-1 beta is induced by lipopolysaccharide, but since neither contain a signal peptide, it is not known how the active protein is secreted into biological fluids. A membrane-bound form of IL-1 may play an important role in both immunological and inflammatory activation (11). In a variety of experimental systems, recombinant-generated IL-1 beta and/or IL-1 alpha have been shown to elicit fever, increase neutrophil phagocytic activity, induce T-lymphocyte proliferation, stimulate release of metalloproteases in synovial tissue and suppress lipid anabolic enzymes (9). Furthermore, recombinant-generated IL-1 mediates changes in expression of hepatic and extrahepatic genes characteristic of the acute phase response (see below).

Cachectin/tumor necrosis factor is a 17.5 kilodalton polypeptide also produced by lipopolysaccharide-activated mononuclear phagocytes. It was independently identified by a group examining the suppression of lipid anabolic enzymes during inflammation (reviewed in 12) and another group studying the spontaneous necrosis of tumors during bacterial infection (reviewed in 13). It is encoded by a gene within the class III region of the major histocompatibility complex on the short arm of chromosome 6. It is adjacent to the locus for its homologue, TNF-beta, or lymphotoxin (14-16). It is structurally distinct from IL-1 and is recognized by a distinct cellular receptor (17). In addition to its effects on lipid metabolism and tumors, cachectin/TNF elicits fever in experimental animals, alters muscle membrane potential, stimulates collagenase and prostaglandin production in cultured synovial cells and has profound effects on vascular endothelium and circulatory integrity. It also mediates changes in expression of several acute phase genes in human hepatoma cells (18).

Interferon beta 2/B-cell differentiation factor BSF-2/hepatocyte stimulating factor (IFN beta 2) is the most recently described mediator of acute phase gene expression. IFN beta 2 is a group of polypeptides derived from a single gene on chromosome 7p21. It was originally described on the basis of its antiviral activity and its capacity to induce proliferation of B-lymphocytes (19). It is now known to elicit many changes in hepatic acute phase gene expression and probably constitutes much of the activity previously termed hepatocyte-stimulating factor (20). It is synthesized in fibroblasts, epithelial cells and mononuclear phagocytes in response to other cytokines (TNF, IL-1, PDGF and IFN beta 1), lipopolysaccharide and phorbol esters (19).

Although probably accounting for a much more limited spectrum of the response, the lymphokines IFN alpha and IFN gamma elicit changes in expression of acute phase genes in hepatocytes (21) and extrahepatic mononuclear phagocytes (22).

Acute Phase Proteins: Complement Factor B and α_1 Proteinase Inhibitor

Within the space available it will not be possible to review our current knowledge of all the acute phase proteins. Instead, these two acute phase proteins will be discussed in detail.

The complement protein factor B is an ~93,000 dalton single chain glycoprotein which forms a part of the alternative complement activating pathway C3 convertase, i.e. it is essential for activation of the complement system via the alternative pathway. Factor B circulates in plasma as a zymogen and is cleaved into amino terminal Ba and carboxy terminal Bb fragments upon activation by factor D. It is encoded by a single gene in class III region of the major histocompatibility complex on chromosome 6p. It is tightly linked to the classical complement activating pathway C3 convertase C2. The C2 gene is ~400 base pairs upstream of the factor B gene (reviewed in 23). Factor B and C2 share other structural and functional similarities.

During inflammation plasma concentrations of factor B increase, but those of its homologue C2 remain unchanged. Likewise, the acute phase monokine IL-1 mediates an increase in synthesis of factor B in human hepatoma cells, mouse fibroblasts transfected with the human factor B gene (4) and in isolated murine hepatocytes (24). In each case the effect involves a pre-translational mechanism of action in that there is a correponding increase in steady state levels of factor B mRNA. The increase in factor B expression has been demonstrated with purified human monocyte IL-1 beta, recombinant murine IL-1 alpha and later recombinant human IL-1 beta (25). In each case there was no change in expression of C2.

Hepatic factor B expression may also be affected by the direct action of at least two other acute phase monokines, TNF-alpha (25) IFN beta 2 (unpublished). TNF-alpha mediates an increase in steady state levels of factor B mRNA and in rates of synthesis of factor B protein. The magnitude of the effect is similar to that of IL-1, but the effect is entirely distinct from that of IL-1. Recombinant human IFN beta 2 mediates an increase in factor B synthesis in HepG2 and Hep3B cells by another specific and direct pathway. The effect of IFN beta 2 is blocked by antibody to the recombinant protein, but not by antibody to recombinant human IL-1; the effect of IL-1 beta is blocked by antibody to recombinant IL-1 beta, but not antibody to recombinant IFN beta 2 (Perlmutter DH and Sehgal P, unpublished).

Factor B expression in hepatocytes may also be regulated by an indirect mechanism involving IL-2. IL-2 is a product of activated lymphocytes which, in turn, stimulates the clonal proliferation of lymphocytes. Patients receiving lymphoblast-derived IL-2 as therapy for disseminated malignancy develop fever, chills, hypotension and changes in serum levels of acute phase reactants (26). Recombinant human IL-2 has no effect on the synthesis of factor B in human hepatoma cells but supernatants of IL-2-activated peripheral blood mononuclear cells mediate a concentration-dependent increase in factor B synthesis and decrease in albumin synthesis (27). This effect is completely neutralized by antibody to IL-1 beta, but not antibody to IL-1 alpha or TNF-alpha. The effect is partially reconstituted in monocytes incubated with both IL-2 and INF-gamma. These results indicate that acute phase regulation of factor B in hepatocytes involves the direct effect of monokines, the direct effect of INF-gamma, and the indirect effect of at least two different lymphokines, IL-2 and INF-gamma.

Factor B expression is also regulated at extrahepatic sites during acute inflammation. Lipopolysaccharide mediates a specific increase in synthesis of factor B in primary cultures of human blood monocytes and bronchoalveolar macrophages (6). Local extrahepatic factor B expression may also be regulated during inflammation in vivo. Adminstration of lipopolysaccharide to mice induces factor B mRNA in lung, kidney, heart, spleen, intestine and peritoneal macrophages, as well as in liver (28). The cell type responsible for extrahepatic factor B gene expression in the latter case is not known, but tissue macrophages are at least in part involved. IL-1 mediates a paracrine or autocrine effect on cultured murine macrophages, increasing synthesis of factor B (29). Factor B expression in human macrophages is also regulated by a direct effect of INF-gamma (22).

Thus activation of macrophages, as might occur under the influence of bacterial lipopolysaccharide or interferon, may directly increase factor B expression at local sites of inflammation or lead to the release of monokines that increase factor B expression in a remote site, liver (Figure 1). Extrahepatic mononuclear phagocyte factor B gene expression may also be modulated by cell-cell interaction, specifically via macrophage membrane IL-1. There is an increase in synthesis of C3 (30) and factor B (Beuscher U, and Colten HR, personal communication) in macrophages incubated on a monolayer of fixed stimulated mouse peritoneal macrophages or incubated on a layer of purified membranes from stimulated mouse peritoneal macrophages.

FIGURE 1

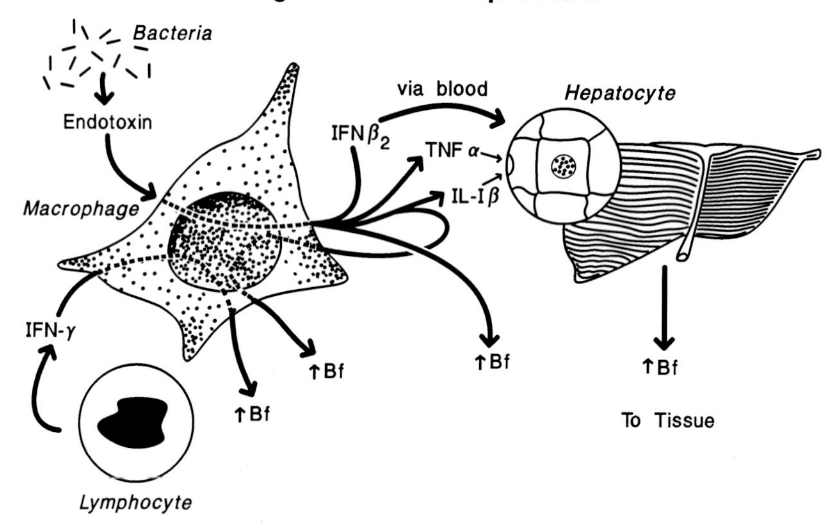

The mechanism of action of IL-1 on hepatic factor B synthesis has been the subject of other studies, now in preliminary stages. First, synthetic peptides have been generated in order to identify the hepatic ligand binding domain of IL-1. Since IL-1 alpha and IL-1 beta have similar biological activity in hepatoma cells, and other tissues, despite only ~26% overall primary sequence homology, a region bearing the highest degree of sequence homology and acrophilic properties was selected for design of the synthetic peptides. It now appears that a peptide of 15 amino acids spanning residues 157-171 of IL-1 beta, antagonizes the effect of IL-1 on synthesis of factor B (Figure 2). For this experiment, HepG2 cells were incubated for 24 hours in control serum-free medium, medium supplemented with peptide 157-171, or medium supplemented with both IL-1 beta and peptide in the specified concentrations. Cells were then labeled for 20 minutes with ^{35}S-methionine and newly synthesized, radiolabeled factor B identified in cell lysates by immunoprecipitation, SDS-PAGE followed by fluorography. The increase in factor B synthesis mediated by IL-1 is almost completely antagonized by peptide at 50-500 mcg/ml (Perlmutter DH, Fok K, Bullock J, Adams SP, unpublished). Further studies with other peptides based on the sequence in this region will be necessary to show antagonism, as well and binding to the IL-1 receptor, at lower concentrations.

FIGURE 2

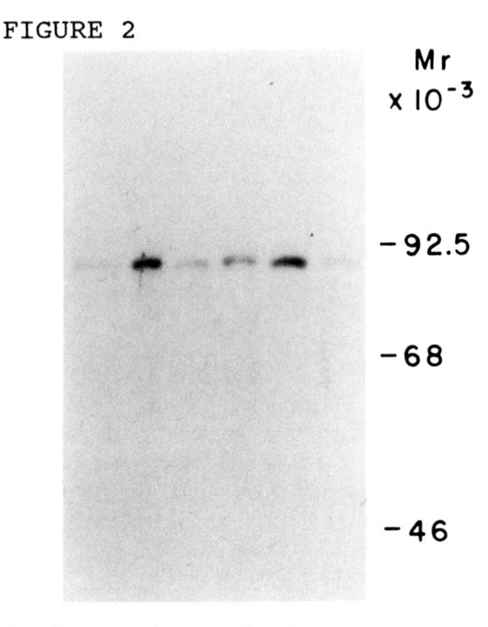

IL-1 (500 pg/ml)	-	+	+	+	+	-
Peptide B (500 mcg/ml)	-	-	+	-	-	+
(50 mcg/ml)	-	-	-	+	-	-
(5 mcg/ml)	-	-	-	-	+	-

Second, structural elements in the upstream flanking region of the factor B gene responsible for the effects of IL-1 on factor B expression have been identified and characterized (3). Plasmid constructs containing various extents of the 5' flanking region of the factor B gene fused to coding region of the reporter gene chloramphenicol acetyl transferase have been expressed in mouse L cells and HepG2 cells. Deletion analysis indicates that the sequence elements responsible for IL-1 regulation are localized 200-600 bases upstream of the initiation site, and are actually localized to the 3' terminus or 3' untranslated region of the C2 gene. These elements are clearly distinct from an interferon consensus sequence at residues -161 to -133. The latter sequence element lies within a fragment which confers INF-gamma modulation to the factor B-CAT fusion gene.

In contrast to factor B, α_1 proteinase inhibitor (α_1 PI) is a human positive acute phase reactant that is not regulated by IL-1, TNF or INF-gamma (25,31). α_1 PI is a 55,000 dalton single chain glycoprotein encoded by an ~12.2 kb gene on chromosome 14 q31-32.3 (32-33). It is a prototype of the serine proteinase inhibitor supergene family that includes α_1-antichymotrypsin (34), anti-thrombin III (35), C1 inhibitor (36-37), α_2-antiplasmin (38), plasminogen activator inhibitor II (39-40), protein C inhibitor (41) and angiotensinogen (42). Each of these inhibitors is able to bind and inactivate a number of serine proteinases in vitro, but has functional specificity for an individual serine proteinase. α_1 PI rapidly inactivates neutrophil elastase and is, therefore, considered an elastase inhibitor under physiologic conditions.

The predominant site of synthesis of plasma α_1 PI is liver (43,44). α_1 PI is synthesized in human hepatoma cells, as well as in primary cultures of human peripheral blood monocytes, bronchoalveolar and breast milk macrophages (45,46). It is now thought that transcription of the α_1 PI gene in macrophages starts ~2 kb upstream from the start site used in hepatocytes

(47,48). Although the same polypeptide is synthesized in the two cell types, a slightly longer mRNA transcript may be present in macrophages depending on alternative post transcriptional splicing of two upstream short open reading frames (48). α_1 PI mRNA has been isolated from a number of tissues in transgenic mice (47,49), but it has not been possible to distinguish whether such α_1 PI mRNA is in ubiquitous tissue macrophages or other cell types. We have recently demonstrated synthesis and secretion of α_1 PI in a human colonic adenocarcinoma cell line (Caco2) that differentiates into a villous enterocyte. α_1 PI mRNA in Caco2 cells and human jejunal epithelium is 1.6 kb corresponding to the size of α_1 PI mRNA in human hepatoma cells but smaller than the ~1.8 kb α_1 PI mRNA in human monocytes and macrophages (Perlmutter DH, Alpers DH, Daniels JD, unpublished). These data indicate that α_1 PI is expressed in small intestine by a cell type other than the macrophage, probably the enterocyte.

Plasma concentrations of α_1 PI increase three- to four-fold during inflammation/tissue injury. The acute phase response of α_1 PI is clearly distinct from that of factor B in that IL-1, TNF and INF-gamma have no effect on synthesis of α_1 PI in HepG2 and Hep3B cells, monocytes or macrophages. Recently we have shown that IFN beta 2 mediates an ~3-fold increase in synthesis of α_1 PI in all these cell types (Perlmutter DH, Sehgal PB, unpublished). This is a direct effect of INF beta 2 in that it is completely neutralized by antibody to recombinant INF beta 2. The characteristics of the regulation of α_1 PI are, in these respects, similar to those of fibrinogen (50).

Expression of α_1 PI in monocytes and macrophages is profoundly influenced by products generated during inflammation. Lipopolysaccharide mediates a 5- to 10-fold increase in synthesis of α_1 PI in mononuclear phagocytes, predominantly increasing the translational efficiency of α_1 PI mRNA (7,51). This is a direct effect of LPS, independent of endogenous mononuclear phagocyte IFN beta 2: the effect of

LPS is not neutralized by antibody to IFN beta 2, but the effect of exogenous IFN beta 2 is neutralized by this antibody (Perlmutter DH, Sehgal PB, unpublished).

Expression of α_1 PI in macrophages is also regulated by the enzyme it inhibits, elastase (52): there is a 4- to 8-fold increase in steady state levels of α_1 PI mRNA and in synthesis of α_1 PI in monocytes incubated with neutrophil elastase. This effect may actually be mediated by an epitope on α_1 PI itself as a consequence of the formation in extracellular fluid of an exogenous elastase-endogenous α_1 PI complex (52). Thus, expression of the acute phase protein α_1 PI during inflammation/tissue injury may involve several cell types, mediators and mechanisms (Figure 3).

FIGURE 3

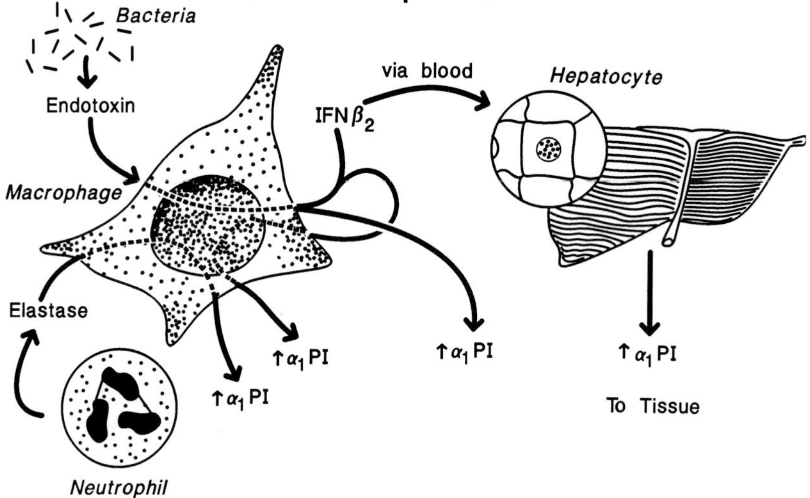

Similarities and differences in the cell types, mediators and mechanisms of regulation involved in the acute phase response of α_1 PI and factor B, as well as other acute phase proteins, may allow us to understand the complex processes invoked during the host response to inflammation.

Acknowledgements

The author is grateful to Dr. Harvey R. Colten for sharing preliminary data and criticism. The work reported here was supported in part by Research Scholar Awards from American Gastroenterological Association/Industry (Merck, Sharp and Dohme) and RJR Nabisco, an Established Investigator Award from the American Association, Monsanto/Washington University Research Agreement and US Public Health Service Grants AI24836, AI24739, HL 37784.

References

1. Tillet WA, Francis T (1930). Serological reactions in pneumonia with a non-protein somatic fraction of pneumococcus. J Exp Med 52:561.
2. Woo P, Korenberg JR, Whitehead AS (1985). Characterization of genomic and complementary DNA sequence of human C reactive protein. J Biol Chem 260:13384.
3. Nonaka M, Gitlin JD, Colten HR (1988). Identification of an IL-1 inducible enhancer region flanking the major histocompatibility complex class III gene, complement factor B (submitted).
4. Perlmutter DH, Goldberger G, Dinarello CA, Mizel SB, Colten HR (1986). Regulation of class III major histocompatibility complex gene products by interleukin-1. Science 232:850.
5. Darlington GJ, Wilson DR, Lachman L (1986). Monocyte-conditioned medium, interleukin-1 and tumor necrosis factor stimulate the acute phase response in human hepatoma cells in vitro. J Cell Biol 103:787.
6. Strunk RC, Whitehead AS, Cole FS (1985). Pretranslational regulation of the synthesis of third component of complement in human mononuclear phagocytes by lipid A portion of lipopolysaccharide. J Clin Invest 76:985.

7. Barbey-Morel C, Pierce JA, Campbell EJ, Perlmutter DH (1987). Lipopolysaccharide modulates the expression of α_1 proteinase inhibitor and other serine proteinase inhibitors in human monocytes and macrophages. J Exp Med 166:1041.
8. Katz Y, Cole FS, Struck RC (1988) Synergism between interferon and lipopolysaccharide for synthesis of factor B, but not C2, in human fibroblasts. J Exp Med 167:1.
9. Dinarello CA (1987). Biology of interleukin 1. FASEB J. 2:108.
10. March CJ, Mosky B, Larsen A, Ceretti DP, Braedt G, Price V, Gillis S, Henney CS, Cronheim CR, Grabstein K, Conlon PJ, Hopp T.P, Cosman D (1985). Cloning, sequence and expression of two distinct human interleukin-1 complementary DNAs. Nature (London) 315:641.
11. Kurt-Jones EA, Beller DI, Mizel SB, Unanue ER (1985). Identification of membrane-associated interleukin 1 in macrophages. Proc Natl Acad Sci USA 82:1204.
12. Beutler B, Cerami A (1986). Cachectin and tumor necrosis factor as two sides of the same biological coin. Nature (London) 320:584.
13. Old LJ (1985). Tumor necrosis factor (TNF) Science 230:630.
14. Spies T, Morton CC, Nedospasov SA, Fiers W, Pious D, Strominger JL (1986). Genes for the tumor necrosis factors alpha and beta are linked to the human major histocompatibility complex. Proc Natl Acad Sci USA 83:8699.
15. Dunham I, Sargent CA, Trowsdale J, Campbell RD (1987) Molecular mapping of the human major histocompatibility complex by pulsed-field gel electrophoresis. Proc Natl Acad Sci USA 84:7237.
16. Carroll MC, Katzman P, Alicot EM, Koller BH, Geraghty DE, Orr HT, Strominger JL, Spies T (1987). Linkage map of the human histocompatibility complex including the tumor necrosis genes. Proc Natl Acad Sci USA 84:8535.

17. Kull FC, Jacobs S, Cuatrecasas P (1985). Cellular receptor for ^{125}I-labelled tumor necrosis factor:specific binding, affinity labelling and relationships to sensitivity. Proc Natl Acad Sci USA 82:5756.
18. Beutler B, Cerami A (1987). Cachectin:more than a tumor necrosis factor. N Engl J Med 316:379.
19. Sehgal PB, May LT (1987). Human β2-interferon. J Interferon Res 7:521.
20. Gauldie J, Richards C, Harnish D, Landsdorp P, Baumann H (1987). Interferon β 2/B-cell stimulatory factor type 2 shares identity with monocyte-derived hepatocyte-stimulating factor and regulates the major acute phase protein response in liver cells. Proc Natl Acad Sci USA 84:7251.
21. Miura N, Prentice H, Schneider PM, Perlmutter DH (1987). Synthesis and regulation of the two human complement C4 genes in transfected murine fibroblasts. J Biol Chem 262:7298.
22. Strunk RC, Cole FS, Perlmutter DH, Colten HR (1985). γ interferon increases expression of class III complement genes C2 and factor B in human monocytes and in murine fibroblasts transfected with human C2 and factor B gene. J Biol Chem 260:15280.
23. Perlmutter DH, Colten HR (1987). Structure and expression of the complement genes. Pharmacol Therap. 34:247.
24. Ramadori G, Sipe JD, Dinarello CA, Mizel SB, Colten HR (1985). Pretranslational modulation of acute phase hepatic protein synthesis by murine recombinant interleukin 1 (IL-1) and purified human IL-1. J Exp Med 162:930.
25. Perlmutter DH, Dinarello CA, Punsal PI, Colten HR (1986). Cachectin/tumor necrosis factor regulates hepatic acute phase gene expression. J Clin Invest 78:1349.
26. Atkins MB, Gould JA, Allegretta M, Dempsey RA, Rudders RA, Parkinson DR, Mier JW (1986). Phase I evaluation of recombinant interleukin 2 in patients with advanced malignant disease. J Clin Oncol 4:1380.

27. Mier JW, Dinarello CA, Atkins MB, Punsal PI, Perlmutter DH (1987). Regulation of hepatic acute phase protein synthesis by products of interleukin 2 (IL-2)-stimulated human peripheral blood mononuclear cells. J Immunol 139:1268.
28. Ramadori G, Sipe JD, Colten HR (1985). Expression and regulation of the murine serum amyloid A (SAA) gene in extrahepatic sites. J Immunol 135:3645.
29. Falus A, Beuscher HU, Auerbach HS, Colten HR (1987). Constitutive and IL-1 regulated murine complement gene expression is strain and tissue specific. J Immunol 138:856.
30. Beuscher HU, Fallon RJ, Colten HR (1987). Macrophage membrane interleukin 1 regulates the expression of acute phase proteins in human hepatoma Hep3B cells. J Immunol 139:1896.
31. Takemura S, Rossing TH, Perlmutter DH (1986). A lymphokine regulates expression of α_1 PI in human monocytes and macrophages. J Clin Invest 77:1207.
32. Pearson SJ, Teton P, George DL, Francke U (1983). Activation of human alpha-1-antitrypsin gene in rat hepatoma x fetal liver cell hybrids depends on the presence of chormosome 14. Somat Cell Genet 9:567.
33. Lai EC, Kao FT, Law ML, Woo SLC (1983). Assignment of the alpha-1-antitrypsin gene and a sequence-related gene to human chromosome 14 by molecular hybridization. Am J Hum Gen 35:385.
34. Chandra T, Stackhouse K, Kidd VJ, Robson KJH, Woo SLC (1983). Sequence homology between human alpha-1-antichymotrypsin and antithrombin III. Biochemistry 22:4996.
35. Hunt LT, Dayhoff MO (1980). A surprising new protein super family containing ovalbumin, antithrombin III and alpha-1-proteinase ihhibitor. Biochem Biophys Res Commun 95:864.

36. Bock SC, Skriver K, Nielsen E, Thogerson H-C, Wiman B, Donaldson DH, Eddy RL, Marinan J, Radziejewska E, Huber R, Shows TB, Magnusson S, (1986). Human C1 inhibitor: primary structure, cDNA cloning and chromosomal localization. Biochemistry 25:4294.
37. Davis AE, Whitehead AS, Harrison RA, Dauphinais A, Burns GAP, Cicardi M, Rosen FS, (1986). Human inhibitor of the first component of complement C1: characterization of cDNA clones and localization of the gene to chromosome 11. Proc Natl Acad Sci USA 83:3161.
38. Holmes WE, Nelles L, Lijnen HR, Collen D (1987). Primary structure of human alpha-2-antiplasmin, a serine protease inhibitor. J Biol Chem 262:1659.
39. Ye RD, Wun T-C, Sadler JE (1987). cDNA cloning and expression in Eschirichia coli of a plasminogen activator inhibitor from human placenta. J Biol Chem 262:3718.
40. Webb AC, Collins KC, Snyder SE, Alexander SJ, Rosenwasser LJ, Eddy RL, Shows TB, Auron PE (1987). Human monocyte Arg-serpin cDNA: Sequence, chromosomal assignment and homology to plasminogen activator-inhibitor. J Exp Med 166:77.
41. Suzuki K, Deyashiki Y, Nishioka J, Kurachi K, Akira M, Yamamoto S, Hashimoto S (1987). Characterization of cDNA for human protein C inhibitor: a new member of plasma serine protease inhibitor super family. J Biol Chem 262:611.
42. Doolittle RF, (1983). Angiotensinogen is related to the antitrypsin-antithrombin-ovalbumin family. Science 222:417.
43. Hood JM, Koep L, Peters RF, Schroter GPJ, Weil R, Redeker AG, Starzl TE (1980). Liver transplantation for advanced liver disease with α-1-antitrypsin deficiency. N Engl J Med 302:272.

44. Alper CA, Raum D, Awdeh ZL, Petersen BH, Taylor PD, Starzl TE (1980). Studies of hepatic synthesis in vivo of plasma proteins including orosomucoid, transferrin, alpha-1-antitrypsin, C8 and factor B. Clin Immunol Immunopathol. 16:84.
45. Perlmutter DH, Cole FS, Kilbridge P, Rossing TH, Colten HR (1985). Expression of the alpha-1-proteinase inhibitor gene in human monocytes and macrophages. Proc Natl Acad Sci USA 82:795.
46. Mornex J-F, Chytil-Weir A, Martinet Y, Courtney M, LeCocq J-P, Crystal RG (1986). Expression of the alpha-1-antitrypsin gene in mononuclear phagocytes and alpha-1-antitrypsin-deficient individuals. J Clin Invest 77:1952.
47. Kelsey GD, Povey S, Bygrave AE, Lovell-Badge RH (1987). Species- and tissue-specific expression of human alpha-1-antitrypsin in transgenic mice. Genes Dev 1:161.
48. Perlino E, Cortese R, Ciliberto G (1987). The human alpha-1-antitrypsin gene is transcribed from two different promoters in macrophages and hepatocytes. EMBO J 6:2767.
49. Sifers RN, Carlson JA, Clift SM, DeMayo FJ, Bullock DW, Woo SLC (1987). Tissue specific expression of the human α-1-antitrypsin gene in transgenic mice. Nucleic Acid Res 15:1459.
50. Evans E, Courtois GM, Kilian PL, Fuller GM, Crabtree GR (1987). Induction of fibrinogen and a subset of acute phase response genes involves a novel monokine which is mimicked by phorbol esters. J Biol Chem 262:10850.
51. Perlmutter DH, Punsal PI (1988). Distinct and additive effects of elastase and endotoxin on α-1-proteinase inhibitor expression in macrophages. J Biol Chem (in press).
52. Perlmutter DH, Travis J, Punsal PI (1988). Elastase regulates the synthesis of its inhibitor, α-1-proteinase inhibitor and exaggerates the defect in homozygous PiZZ α-1-PI deficiency. J Clin Invest (in press).

Stress-Induced Proteins, pages 275–285

STRESS-INDUCED PROTEINS AS ANTIGENS IN INFECTIOUS DISEASES

Douglas Young, Raju Lathigra, and Angela Mehlert

MRC Tuberculosis and Related Infections Unit
Royal Postgraduate Medical School, Hammersmith Hospital,
London, W12 0HS, U.K.

ABSTRACT Analysis of components recognised during the immune response to infection has demonstrated that a variety of bacterial and parasitic pathogens express major antigens which are members of conserved stress-induced protein families. Among the hypotheses which can be proposed to account for the apparent immunodominance of such proteins, it is attractive to speculate that the process of adaptation to the host environment during infection may involve the induction of stress protein synthesis. The extensive sequence homology between the corresponding pathogen and host cell proteins suggests potential involvement of such structures in the induction of tolerance or in the generation of autoimmune pathology. Detailed analysis of the regulation of the stress response and of the immune response to stress protein determinants will be important in understanding the role of these proteins in infection and immunity.

INTRODUCTION

Stimulation of the immune system during chronic infection is a complex event which can involve responses which are apparently detrimental to the host, in addition to the responses which mediate protective immunity. During infection with *Mycobacterium leprae*, for example, activation of the cellular immune system can play a role in the neuropathology associated with tuberculoid leprosy, while induction of immune tolerance can lead to the unrestricted bacterial growth seen in the lepromatous form of the disease (1). Production of vaccines to combat such infections will depend on the ability to selectively stimulate protective immune functions whilst avoiding simultaneous stimulation of tolerance or immunopathology. In order to dissect the immune response in this way, attempts are being made to isolate individual antigens from pathogens and to carry out a detailed analysis of their interaction with components of the immune system. A goal of such research would be to develop "subunit vaccines" capable of inducing protective immune responses. An unexpected outcome of this approach has been the finding that many antigens isolated from a diverse range of pathogens are members of stress-induced protein families. In this paper we discuss the implications of these observations, particularly from the perspective of our own interest in mycobacterial infection.

ANTIGENS WHICH BELONG TO STRESS PROTEIN FAMILIES

The hsp 70 Family

The hsp 70 gene family contains the major heat inducible proteins which are found in all living organisms with a high degree of sequence conservation extending from bacteria to mammalian cells (2).

Analysis of major mycobacterial antigens using murine monoclonal antibodies has identified antigenically related proteins from M.tuberculosis and M.leprae with molecular weights of 71 and 70kD respectively (3,4). The 71/70kD antigen is also involved in the recognition of mycobacteria by human T lymphocytes during infection (5). This antigen was identified as a member of the hsp 70 gene family by sequence analysis of the cloned gene (6). Labelling studies during heat shock, and the biochemical and antigenic properties of the purified mycobacterial 71kD protein confirmed that it is homologous with other hsp 70 proteins (7). Polyclonal antisera raised to the mycobacterial antigen also bind to DnaK (the E.coli member of the hsp 70 gene family) and to hsp 70 proteins isolated from human cells (7).

Two reports have described the characterisation of a 75kD antigen from the malarial parasite, Plasmodium falciparum, which was isolated by screening cDNA expression libraries using immune sera (8,9). Sequence analysis has identified this antigen as a member of the hsp 70 gene family.

Similar techniques resulted in the cloning of an antigen from Schistosoma mansoni which was also subsequently found to belong to the hsp 70 family (10). Further examples of hsp 70 proteins being recognised by the immune system occur in the case of the filarial parasite (Brugia malayi) (11), Trypanosoma cruzi (12) and Leishmania major (13).

The GroEL Family

The products of the groEL and adjacent groES genes are major components in the heat shock response of E.coli (2). Until recently the GroEL protein was thought to be restricted to bacteria, but it is now clear that a family of proteins with conserved structural elements and antigenic determinants extends to chloroplasts and mitochondria of higher organisms (14,15).

The 65kD protein of mycobacteria is an immunodominant antigen in murine antibody and cell mediated immune responses, as assessed by the frequency of monoclonal antibody production and by limiting dilution analysis of T lymphocytes (3,4,16). Responses to this antigen are also frequently detected during analysis of human T cell recognition of mycobacteria (17,18). The genes for the 65kD proteins of M.leprae, M.tuberculosis and M.bovis BCG have been cloned and sequenced (19-21) and many of the determinants involved in antibody and T cell recognition have been identified and reproduced using synthetic peptides (18,19). It is now clear that this 65kD antigen belongs to the GroEL family and that the mycobacterial and E.coli proteins have a high degree of sequence conservation and share many antigenic determinants (6). The 65kD antigen is also a major heat shock protein (22), but a mycobacterial counterpart of the GroES protein has not yet been identified by antibody or sequence analysis.

An antigen of Coxiella burnetii (the causative agent of Q fever) which was isolated from a genomic DNA expression library by screening with antisera, has also been shown to belong to the GroEL family (23). In this case a low molecular weight protein having approximately 50% sequence homology with E.coli GroES was found. In contrast to the GroEL homologue, the Coxiella GroES homologue was not found to be strongly antigenic (23). A protein with structural and antigenic features consistent with its identification as a member of the GroEL family is a prominent target of the antibody response to infection with Legionella (24). GroEL antigens from

Treponema pallidum and *Borrelia burgdorferi* have also been cloned and shown to correspond to a previously described "common bacterial antigen" (25,26).

Although it is clear that the eukaryotic GroEL homologues share antigenic cross-reactivity with the bacterial proteins, there are at present no reports of their immune recognition during parasite infections.

Other Stress-induced Proteins

An 18kD antigen from *M.leprae* has been identified by its interaction with mouse monoclonal antibodies (3) and with human T lymphocytes (27). The gene for this antigen has been cloned and sequenced (28). Close examination of the sequence of the 18kD protein indicated that this antigen may belong to the loosely conserved family of low molecular weight heat shock proteins (6). A protein antigen from *Schistosoma mansoni* which belongs to the low molecular weight heat-shock protein family has also been reported (29). The extent of the sequence conservation in these two instances is very much less than that observed for the hsp 70 and GroEL families.

Summary

Table 1 summarises examples of proteins which were isolated from bacteria and parasites on the basis of their immunological activity and which have now been identified as members of stress-induced protein families. This list has increased rapidly over the last year and it seems likely that it will continue to grow.

THE IMMUNE RESPONSE TO STRESS PROTEINS

In the light of the observation that proteins belonging to the stress-induced protein families are often strongly antigenic, it is interesting to speculate as to why this class of proteins should be so attractive to the immune system. Common features of these proteins - stress induction, role in protein folding and assembly, sequence conservation - could all be considered as potential factors contributing to their immunogenicity.

Table 1. Antigens which belong to stress protein families.

Organism	Disease	Stress Protein Family			References
		hsp 70	GroEL	low mol.wt.	
Bacteria					
Mycobacteria	leprosy tuberculosis	+	+	+	(6)
Coxiella	Q fever		+		(23)
Legionella	legionaires' disease		+		(24)
Treponema	syphilis		+		(25)
Borrelia	Lyme disease		+		(26)
Parasites					
Plasmodia	malaria	+			(8,9)
Schistosomes	schistosomiasis	+		+	(10,29)
Brugia	filariasis	+			(11)
Trypanosomes	trypanosomiasis	+			(12)
Leishmania	leishmaniasis	+			(13)

Induction of stress proteins during infection

Members of the hsp 70 gene family are induced as part of the parasite differentiation process which occurs during adaptation to life within the infected host (30). It is possible that this induction leads to increased levels of such proteins being made available to the immune system during establishment of the host-parasite relationship and that this may lend a particularly enhanced antigenic profile to these structures. An analogous situation could be postulated for intracellular bacteria - such as mycobacteria - which have to adapt to life within the hostile environment of phagocytic host cells. *Salmonella typhimurium* mutants carrying a defect in their ability to up-regulate proteins involved in the response to oxidative stress are unable to survive inside macrophages (31), and it can be postulated that stress proteins are an essential protection for bacteria which have to survive the microbicidal mechanisms of the immune system.

The concept that up-regulation during infection may enhance immune recognition of a particular protein could be of importance in understanding the differential immune response to live and to dead organisms. Expression of an antigen under the control of a stress promoter may, for example, be an interesting strategy in the design of live vectors for vaccination. While stress regulation may contribute to immunogenicity, it is clear that other structural properties of the proteins also influence their interaction with the immune system. GroEL homologues induce a strong antibody response when presented as purified proteins or as crude bacterial extracts, while GroES (which is co-regulated with GroEL) is poorly immunogenic (unpublished observations).

Structural features of stress proteins

A common feature of stress protein families is their participation in multi-molecular structures either by self-aggregation or by interaction with other proteins or intracellular structures. Perhaps a property of this kind renders these proteins particularly effective in stimulation of antigen processing or in some stage of lymphoctye activation. Examples of other aggregated proteins which are particularly effective in the induction of immune responses include the large T antigen of the SV40 virus and the hepatitis B core antigen (32).

Alternatively the ability of stress proteins to associate with particular intracellular structures may play a role in their immunogenicity. The ability to form a complex with the p53 oncogene product, for example, is shared by eukaryotic and prokaryotic hsp 70 proteins (33) and is presumably a function associated with the conserved structural features of these proteins. Perhaps complex formation with specific proteins within the antigen processing cell has a role in preferential immune recognition of stress proteins.

Stress proteins, tolerance and autoimmunity

Since members of the stress-induced protein families are common constituents of all organisms, it is clear that the immune system will be constantly exposed to such structures from gut flora and dietary sources, in addition to any infections. It is possible that the intensity of the immune response to these antigens simply reflects the frequency with which thay are encountered. However, the antigens produced by the pathogenic organisms have a remarkable degree of sequence conservation with the corresponding proteins of the host cells, and it would be anticipated that at least part of the antigens should be seen as "self" proteins by the immune system. Presentation of "self-like" antigens could induce tolerance towards the pathogen, allowing it to escape from immune surveillance during the early stages of infection. Breakdown of this tolerance which resulted in the recognition of conserved portions of the "foreign" stress proteins would constitute an autoimmune response, with potentially pathogenic consequences.

The presence of a population of potentially autoreactive lymphocytes recognising stress protein determinants may not in itself be sufficient to cause a disease. In the event of a subsequent lesion resulting in elevated expression of the host protein - in response to a stress stimulus viral infection, for example - stimulation of the autoreactive cells by the self antigen could be a cause of tissue damage. A model of this type could be used to analyse the mechanisms involved in autoimmune diseases, such as rheumatoid arthritis, as well as to explain the immunopathology associated with chronic infections.

Data indicating a possible role for stress proteins as autoantigens has recently become available. Using a rat model of adjuvant induced arthritis it has been shown that disease can be transferred by a population of autoreactive T lymphocytes directed to a member of the GroEL family (34). We are currently carrying out a detailed analysis of human T cell recognition of antigenic determinants on the GroEL homologues in order to assess the potential role of this family of proteins in human autoimmune dieases. Autoantibodies to hsp 90 have been found in sera from patients with systemic lupus erythematosus (SLE) (35), and antibodies to hsp 70 are significantly elevated in patients with SLE and with rheumatoid arthritis (36).

CONCLUSIONS

Antigens which belong to the stress-induced protein families are clearly recognised in a variety of infectious diseases, but could these proteins be potential candidates for use as subunit vaccines? Stress proteins are not generally expressed on the surface of cells and it is unlikely therefore that they will provide useful targets for antibody/complement mediated killing. T cell recognition of stress proteins, as illustrated in the case of the mycobacterial antigens, suggests that they may have a role in providing help for antibody responses to other proteins and could be important in the stimulation of macrophage activation. The possible involvement of stress proteins in induction of tolerance or autoimmunity however, suggests that manipulation of the immune response to these antigens could have detrimental as well as beneficial consequences.

REFERENCES

1. Bloom BR, Godal T (1983). Selective primary health care: stategies for control of disease in the developing world. Leprosy. Rev Infect Dis 5:765.
2. Lindquist S (1986). The heat-shock response. Ann Rev Biochem 55:1151.
3. Engers et al. (1985). Results of a WHO sponsored workshop on monoclonal antibodies to Mycobacterium leprae. Infect Immun 48:603.
4. Engers et al. (1986). Results of a WHO sponsored workshop to characterise antigens recognised by mycobacteria-specific monoclonal antibodies. Infect Immun 51:718.

5. Britton WJ, Hellqvist L, Basten A, Inglis AS (1986). Immunoreactivity of a 70kD protein purified from Mycobacterium bovis BCG by monoclonal antibody affinity chromatography. J Exp Med 164:695.
6. Young DB, Lathigra R, Hendrix R, Sweetser D, Young RA (1988). Stress proteins are immune targets in leprosy and tuberculosis. Proc Natl Acad Sci USA, in press.
7. Mehlert A, Lamb J, Young DB (1988). Analysis of stress-related proteins involved in the immune response to mycobacterial infection. Biochem Soc Trans, in press.
8. Bianco AE, Favaloro JM, Burkof TR, Culvenor JG, Crewther PE, Brown GV, Anders RF, Coppel RL, Kemp DJ (1986). A repetitive antigen of Plasmodium falciparum that is homologous to heat shock protein 70 of Drosophila melanogaster. Proc Natl Acad Sci USA 83:8713.
9. Ardeshir F, Flint JE, Richman J, Reese RT (1987). A 75kD merozoite surface protein of Plasmodium falciparum which is related to the 70 kD heat-shock proteins. EMBO J 6:493.
10. Hedstrom R, Culpepper PJ, Harrison RA, Agabian N, Newport G (1987). A major immunogen in Schistosoma mansoni infections is homologous to the heat-shock protein hsp70. J Exp Med 165:1430.
11. Selkirk ME, Rutherford PJ, Denham DA, Partono F, Maizels RM (1988). Biochem Soc Symp, in press.
12. Engman DM, Kirchoff LV, Henkle K, Donelson JE (1988). A novel hsp70 cognate in trypanosomes. J Cell Biochem 12D:290.
13. Smith DF, Searle S, Campo AJR, Coulson RMR, Ready PD (1988). A multigene family in Leishmania major with homology to eukaryotic heat shock protein 70 genes. J Cell Biochem 12D:296.
14. Ellis J (1987). Proteins as molecular chaperones. Nature 328:378.
15. McMullin TW, Hallberg RL (1988). A highly evolutionarily conserved mitochondrial protein is structurally related to the protein encoded by the E.coli groEL gene. Mol Cell Biol 8:371.
16. Kaufman SHE, Vath U, Thole JER, van Embden JDA, Emmrich F (1987). Enumeration of T cells reactive with Mycobacterium tuberculosis organisms and specific for the recombinant 64kDa protein. Eur J Immunol 17:351.
17. Emmrich F, Thole JER, van Embden JDA, Kaufman SHE

(1985). A recombinant 64 kDa protein of Mycobacterium bovis BCG specifically stimulates human T clones reactive to mycobacterial antigens. J Exp Med 163:1024.

18. Lamb JR, Ivanyi J. Rees ADM, Rothbard JB, Howland K, Young RA, Young DB (1987). Mapping of T cell epitopes using recombinant antigens and synthetic peptides. EMBO J 6:1245.
19. Mehra V, Sweetser D, Young RA (1986). Efficient mapping of protein antigenic determinants. Proc Natl Acad Sci USA 83:7013.
20. Shinnick TM (1987). The 65-kD antigen of Mycobacterium tuberculosis. J Bacteriol 169:1080.
21. Thole JER, Keulen WJ, Kolk AHD, Groothius DG, Berwald LG, Tiesjema RH, van Embden JDA (1987). Characterization, sequence determination, and immunogenicity of a 64-kD protein of Mycobacterium bovis BCG expressed in E.coli K-12. Infect Immun 55:1466.
22. Shinnick TM, Vodkin MH, Williams JC (1988). The Mycobacterium tuberculosis 65-kD antigen is a heat shock protein which corresponds to common antigen and to E.coli GroEL protein. Infect Immun 56:446.
23. Vodkin MH, Williams JC (1988). A heat shock operon in Coxiella burnetii produces a major antigen homologous to a protein in both mycobacteria and E.coli. J Bacteriol 170:1227.
24. Plikaytis BB, Carlone GM, Pau C-P, Wilkinson HW (1987). Purified 60kD Legionella protein antigen with Legionella-specific and nonspecific epitopes. J Clin Microbiol 25:2080.
25. Hindersson P, Knudson JD, Axelsen NH (1987). Cloning and expression of Treponema pallidum common antigen (Tp-4) in E.coli K-12. J Gen Microbiol 133:587.
26. Hansen K, Fjordang H, Pedersen NS, Hinderssen P (1988). Borrelia burgdorferi expresses an immunodominant 60kD antigen common to a wide range of bacteria: immunochemical characterisation and isolation of the gene. submitted for publication.
27. Mustafa AS, Gill HK, Nerland A, Britton WJ, Mehra V, Bloom BR, Young RA, Godal T (1986). Human T cell clones recognise a major M.leprae antigen expressed in E.coli. Nature 319:63.
28. Booth RJ, Harris DP, Love JM, Watson JD (1988).

Antigenic proteins of M.leprae: complete sequence of the gene for the 18kD protein. J Immunol 140:597.
29. Nene V, Dunne DW, Johnson KS, Taylor DW, Cordingley JS (1986). Sequence and expression of a major egg protein from Schistosoma mansoni. Homologies to heat shock proteins and alpha crystallins. Mol Biochem Parasitol 21:179.
30. Van der Ploeg LK, Giannini SH, Cantor CR (1985). Heat shock genes: regulatory role for differentiation in parasitic protozoa. Science 228:1443.
31. Christman MF, Morgan RW, Jacobson FS, Ames BN (1985). Positive control of a regulon for defences against oxidative stress and some heat shock proteins in Salmonella typhimurium. Cell 41:753.
32. Clarke BE, Newton SE, Carroll AR, Francis MJ, Appleyard G, Syred AD, Highfield PE, Rowlands DJ, Brown F (1987). Improved immunogenicity of a peptide epitope after fusion to hepatitis B core protein. Nature 330:381.
33. Clarke CF, Cherg K, Frey AB, Stein R, Hinds PW, Levine AJ, (1988). Purification of complexes of nuclear oncogene p53 with rat and E.coli heat shock proteins: in vitro dissociation of hsc70 and dnaK from murine p53 by ATP. Mol Cell Biol 8:1206.
34. Van Eden W, Thole JER, van der Zee R, Noordzij A, van Embden JDA, Hensen EJ, Cohen IR (1988). Cloning of the mycobacterial epitope recognised by T lymphocytes in adjuvant arthritis. Nature 331:171.
35. Minot S, Koyasu S, Yahara I, Winfield J (1988). Autoantibodies to the heat shock protein hsp90 in systemic lupus erythematosus. J Clin Invest 81:106.
36. Tsoulfa G, Rook G, Mehlert A, Young DB, Isenberg D, Lydyard P. manuscript in preparation.

Index